T0406079

Welded High Strength Steel Structures

Welded High Strength Steel Structures

Welding Effects and Fatigue Performance

Jin Jiang

Author

Prof. Jin Jiang
School of Civil and Environmental Engineering
Shantou University
China

Cover Images: © Stefano Marsella/Unsplash, © Karan Bhatia/Unsplash

Library of Congress Card No.:
applied for

British Library Cataloguing-in-Publication Data
A catalogue record for this book is available from the British Library.

Bibliographic information published by the Deutsche Nationalbibliothek The Deutsche Nationalbibliothek lists this publication in the Deutsche Nationalbibliografie; detailed bibliographic data are available on the Internet at <http://dnb.d-nb.de>.

Print ISBN: 978-3-527-34726-1
ePDF ISBN: 978-3-527-34727-8
ePub ISBN: 978-3-527-83184-5
oBook ISBN: 978-3-527-83185-2

Cover Design Wiley

Typesetting Set in 9.5/12.5pt STIXTwoText by Integra Software Services Pvt. Ltd, Pondicherry, India

Printing and Binding CPI Antony Rowe

Printed on acid-free paper

Contents

List of Figures *ix*
List of Tables *xvii*
Preface *xix*
Notation *xxi*

1 Introduction *1*
1.1 Research Background *1*
1.2 Objectives and Scope *4*
1.3 Contributions and Originality *6*
1.4 Organization *7*

2 Literature Review *9*
2.1 High-Strength Steel *9*
2.1.1 Overview *9*
2.1.2 Delivery Condition of HSS *10*
2.1.3 Fatigue and Fracture of HSS *11*
2.1.4 Codes and Standards of HSS Application *12*
2.2 Welding and Residual Stress *13*
2.2.1 Overview of Arc Welding *13*
2.2.2 Weldability of Steel *15*
2.2.3 Phase Transformation and Other Phenomena in Welding Procedures *16*
2.2.4 The Formation of Residual Stress *20*
2.2.4.1 Origin and Types of Residual Stress *20*
2.2.4.2 Generation of Welding Residual Stress *22*
2.2.5 Residual Stress Investigation Techniques *24*
2.2.5.1 Experimental Investigation *24*
2.2.5.2 Numerical Modeling *26*
2.2.6 Exploration on Residual Stress Effects *28*
2.3 Fatigue Analysis of Tubular Joints *31*
2.3.1 Classification and Parameters of Tubular Joints *31*
2.3.2 Stress Analysis of Intact Tubular Joints *32*

3 Experimental Investigation of Residual Stress for High-Strength Steel Plate-to-Plate Joints *37*
3.1 Introduction *37*
3.2 The Hole-Drilling Method and Specimen Details *38*
3.2.1 The ASTM Hole-Drilling Method *38*
3.2.2 Specimen Specifications *39*
3.2.3 Welding Specifications *41*
3.3 Residual Stress Investigation *43*
3.3.1 Setup and Modification of the Hole-Drilling Guide *43*
3.3.2 Strain Gauge Locations *44*
3.3.3 Calibration Test for Residual Stress Measurement *45*
3.3.4 Residual Stress Measurement Procedure *46*
3.3.5 Cutting of Brace Plate *47*
3.4 Experimental Results *47*
3.4.1 Distribution of Residual Stress Along the Weld Toe *49*
3.4.2 The Effects of Preheating *49*
3.4.3 The Effects of Joint Angle *50*
3.4.4 The Effects of Plate Thickness *51*
3.4.5 The Effects of Brace Plate Cutting *52*
3.5 Static Tensile Testing *54*
3.5.1 Testing Rig *54*
3.5.2 Strain Gauge Locations *54*
3.5.3 Testing Procedure *54*
3.5.4 Testing Results *56*
3.6 The Influence of Residual Stress on SCF Value *57*
3.6.1 Analysis Method *57*
3.6.2 Results and Conclusions *58*
3.7 Conclusion and Summary *60*

4 Numerical Study of Residual Stress for High-Strength Steel Plate-to-Plate Joints *63*
4.1 Introduction *63*
4.2 Modeling Procedure and Results for 2D Models *64*
4.2.1 Overview *64*
4.2.2 Lumped Technique *64*
4.2.3 Weld Filler Addition Technique *67*
4.2.4 Heat Transfer Analysis *68*
4.2.5 Mechanical Analysis *70*
4.2.6 Model Validation and Results *71*
4.2.6.1 Model Validation *71*
4.2.6.2 Numerical Modeling Results *72*
4.3 Modelling Procedure and Results for 3D Models *76*
4.3.1 Overview *76*
4.3.2 Heat Source Model in 3D Analysis *77*
4.3.3 Modeling for the Weld Filler Adding Process *78*
4.3.4 Modeling Validation *80*

4.3.5 Modeling Results *81*
4.3.5.1 Ambient Temperature Joint *81*
4.3.5.2 Preheating Joint *83*
4.3.5.3 Comparison Between Ambient Temperature and Preheated Joints *84*
4.4 Parametric Study *87*
4.4.1 Effect of Boundary Condition *91*
4.4.2 Effect of Preheating Temperature *91*
4.4.3 Effect of Using Different Lumps *93*
4.4.4 Effect of Welding Speed *94*
4.4.5 Effect of Welding Sequence *95*
4.5 Conclusions *96*

5 Experimental Investigation of Residual Stress for Welded Box High-Strength Steel T-Joints *99*
5.1 Introduction *99*
5.2 Experimental Investigation *100*
5.2.1 Material Properties *100*
5.2.2 Specimen Fabrication *101*
5.2.2.1 Overview of the Welding Design *101*
5.2.2.2 Fabrication of Box Sections *104*
5.2.2.3 Fabrication of Joint Intersection *104*
5.2.3 Residual Stress Test Setup and Procedure *105*
5.2.4 Strain Gauge Schemes for Residual Stress Measurement *106*
5.2.5 Computation of Residual Stress *107*
5.3 Testing Results *109*
5.3.1 Preheated Specimen *109*
5.3.2 Ambient Temperature Specimen *114*
5.4 Analyses and Discussion *118*
5.4.1 Preheating Effect *118*
5.4.2 Chord Edge Effect *120*
5.4.3 Corner Effect *120*
5.4.4 Stress Variation in Depths *121*
5.5 Conclusions *123*

6 Numerical Study of Residual Stress for Welded High-Strength Steel Box T/Y-Joints *125*
6.1 Introduction *125*
6.2 Modeling Procedure *126*
6.2.1 Overview *126*
6.2.2 Heat Source Modeling *129*
6.2.3 Thermal Interactions *129*
6.2.4 Arc Touch Movement *130*
6.2.5 Modeling Summary *130*
6.3 Modeling of Pure Heat Transfer *132*
6.4 Fully Coupled Residual Stress Analysis *136*
6.4.1 Modeling Validation *136*

6.4.2 Modeling Results *137*
6.4.2.1 Temperature History *137*
6.4.2.2 Residual Stress *138*
6.5 Parametric Study *141*
6.5.1 Range of the Modeling *141*
6.5.2 Variation of the Residual Stress with Respect to Joint Angle *142*
6.5.2.1 Variation of the Residual Stress with Respect to Joint Angle and Welding Starting Location *142*
6.5.2.2 Variation of the Residual Stress with Respect to Joint Angle and Preheating Temperature *143*
6.5.3 Variation of the Residual Stress with Respect to b/c (Ratio of Brace Width to Chord Width) *145*
6.5.3.1 Variation of the Residual Stress with b/c and Preheating Temperature *145*
6.5.3.2 Variation of the Residual Stress with b/c and Welding Starting Location *146*
6.5.4 Variation of the Residual Stress with Respect to Welding Speed *147*
6.5.4.1 Variation of the Residual Stress with Welding Speed and Preheating Temperature *147*
6.5.4.2 Variation of the Residual Stress with Welding Speed and b/c *148*
6.6 Conclusions *149*

7 Stress Concentration Factor of Welded Box High-Strength Steel T-Joint *153*
7.1 Introduction *153*
7.2 Test Setup and Specimens *154*
7.3 Strain Gauge Schemes *156*
7.4 Test Procedure *158*
7.5 Test Results *159*
7.6 Comparision of Test Results with CIDECT Guide *161*
7.7 Effect of Residual Stress on SCF *162*
7.8 Conclusion *163*

8 Conclusion and Recommendation *165*
8.1 Introduction *165*
8.2 Conclusions *166*
8.2.1 Experimental Studies *166*
8.2.2 Numerical Modeling *167*
8.3 Recommendations for Future Research Work *168*

Summary *169*
Appendix 1 *171*
Appendix 2 *175*
Appendix 3 *181*
References *195*
Index *201*

List of Figures

1.1 Sony center in Berlin. 2
1.2 Application of HSS in Sony center in Berlin. 2
1.3 Historical development of production processes for rolled steel products (Sedlacek and Muller 2005). 3
1.4 Reduction of wall thickness and weight with increasing strength of steel (Sedlacek and Muller 2005). 3
1.5 Technical flow of the book. 5
2.1 Comparison of different steel delivery conditions. 10
2.2 Steel microstructures for different delivery conditions. 11
2.3 Comparison of yield stress for different delivery conditions. 11
2.4 The nature of welding. 14
2.5 Mechanism of fusion welding. 15
2.6 Relationship between yield stress and $C_{\text{equiv.}}$ 16
2.7 Phase transformation in the welding. 17
2.8 CCT diagrams of HSS. 18
2.9 Columnar grain structure of weld. 18
2.10 Criterion of hot crack in the weld. 19
2.11 Welding and mechanical properties of steel. 20
2.12 Residual stress variation along heat moving path. 21
2.13 Three types of residual stress. 22
2.14 Heat transfer analysis in the welding process. 23
2.15 Couplings in residual stress formation. 26
2.16 Residual stress relaxation after applying tensile stress. 29
2.17 Impact of residual stress on K value. 31
2.18 Common rectangular tubular joint configurations. 32
2.19 Geometrical parameters for tubular joints. 32
2.20 Typical S–N curves for fatigue design of CHS and RHS joints. 33
2.21 Definition of hot-spot stress. 34
3.1 Schematic diagram of strain gauge for residual stress measurement. 39
3.2 Typical welding profile of plate-to-plate joint (for $\theta = 90°$ and 135°). 41
3.3 Preheating area. 42
3.4 Preheating process. 42
3.5 The number of welding passes and welding sequences adopted for joints with different thickness and intersection angles. 43

3.6 The RS-milling guide for residual stress measurement. 44
3.7 Strain gauge locations on chord plate for residual stress measurement (all dimensions in mm). 45
3.8 Calibration test for residual stress coefficients. 46
3.9 A T-joint before cutting of the brace plate. 47
3.10 A T-joint after cutting of the brace plate. 48
3.11 Residual stresses distribution along the transverse direction (y axis, gauges B, B1, B2, and B3) for 90° joints. 49
3.12 Residual stresses distribution along the transverse direction (y axis, gauges B, B1, and B2) for 135° joints. 50
3.13 Residual stresses distribution along the transverse direction (y axis) for joints welded at ambient temperature. 51
3.14 Residual stresses distribution along the transverse direction (y axis) for joints with preheating. 52
3.15 Effects of brace plate cutting for 90° joints welded at ambient temperature. 53
3.16 Effects of brace cutting for 90° joints with preheating. 54
3.17 Strain gauges scheme for SCF measurement. 55
3.18 Plane view of scheme of strain gauges for static testing. 55
3.19 Strain gauge locations in specimen for static tensile testing. 55
3.20 Assembly of the specimen and supporting joints. 56
3.21 Fixing of the specimen and the supporting joints in the testing machine. 57
3.22 *RSF* of 135° joints under different nominal stresses. 59
3.23 *RSF* of 90° joints under different nominal stresses. 60
4.1 The flow chart of the modeling procedure. 65
4.2 Discretization of the plate-to-plate joints (Light Gray elements: activated; Dark Gray elements: deactivated.) 66
4.3 Modeling techniques for adding of weld filler. 67
4.4 Thermal properties used in the modeling. 69
4.5 Mechanical properties from EC3 and testing. 70
4.6 Comparison of modeling and testing results for $\theta = 90°$ joints with preheating. 72
4.7 Comparison of modeling and testing results for $\theta = 90°$ joints welded at ambient temperature. 72
4.8 Comparison of modeling and testing results for $\theta = 135°$ joints with preheating. 73
4.9 Comparison of modeling and testing results for $\theta = 135°$ joints welded at ambient temperature. 73
4.10 The temperature distributions at different times ($\theta = 135°$, $t_1 = 12$ mm, welded at ambient temperature) (a) t = 1.0s, (b) t = 58.7s, (c) t = 117.4s, (d) t = 176.1s, (e) t = 300s, (f) t=2500s. 74
4.11 The residual stress near weld in joints ($\theta = 135°$, $t_1 = 12$ mm). 74
4.12 Relationship between transverse residual stress and distance from the weld toe ($\theta = 135°$, $t_1 = 12$ mm, preheated). 75
4.13 Relationship between transverse residual stress and distance from the weld toe ($\theta = 135°$, $t_1 = 12$ mm, ambient temperature). 75

4.14 The average cooling rate at selected points for preheating and ambient temperature cases ($\theta = 135°$, $t_1 = 12\,\text{mm}$). 76
4.15 Double ellipsoidal heat source model. 77
4.16 Welding direction in the selected HSS plate-to-plate joint ($\theta = 135°$, $t_1 = 12\,\text{mm}$, ambient temperature and $\theta = 135°$, $t_1 = 12\,\text{mm}$, preheating, *LHS:* left hand side, *RHS:* right hand side). 78
4.17 Modeling of welding direction along joint width. 78
4.18 Meshing for the HSS plate-to-plate joint. 79
4.19 Element birth and death technique used in the modeling (welding direction, *case 1*). 80
4.20 Comparison for 2D and 3D modeling and test results ($\theta = 135°$, $t_1 = 12\,\text{mm}$, ambient temperature). 80
4.21 Comparison for 2D and 3D modeling and test results ($\theta = 135°$, $t_1 = 12\,\text{mm}$, preheating). 81
4.22 The transverse residual stress profile for the ambient temperature specimen (*LHS:* left-hand side, *RHS:* right-hand side) 82
4.23 Transverse residual stress variation at different locations ($\theta = 135°$, $t_1 = 12\,\text{mm}$, ambient temperature, welding direction: case 1). 82
4.24 Transverse residual stress variation at different depths ($\theta = 135°$, $t_1 = 12\,\text{mm}$, ambient temperature, welding direction: case 1). 83
4.25 The transverse residual stress profile for preheated specimen at the chord plate (*LHS:* left-hand side, *RHS:* right-hand side). 84
4.26 Transverse residual stress variation at different locations ($\theta = 135°$, $t_1 = 12\,\text{mm}$, preheated, welding direction: *case 1*). 85
4.27 Transverse residual stress variation at different depths ($\theta = 135°$, $t_1 = 12\,\text{mm}$, preheated, welding direction: *case 1*). 85
4.28 The von Mises residual stress in the middle of chord width ($\theta = 135°$, $t_1 = 12\,\text{mm}$, ambient temperature, welding direction: *case 1*). 85
4.29 The von Mises residual stress in the middle of chord width ($\theta = 135°$, $t_1 = 12\,\text{mm}$, preheating, welding direction: *case 1*). 86
4.30 The transverse residual stress at the middle of chord width ($\theta = 135°$, $t_1 = 12\,\text{mm}$, ambient temperature, welding direction: *case 1*). 86
4.31 The transverse residual stress at the middle of chord width ($\theta = 135°$, t_1=12 mm, preheating, welding direction: *case 1*). 87
4.32 Transverse residual stress variation at different locations. 87
4.33 Three boundary conditions included in the modeling. 90
4.34 Different lumping scheme in the parametric study. 90
4.35 Cases for different weld sequence in the parametric study. 91
4.36 Comparison of transverse residual stress under different boundary conditions. 91
4.37 Comparison of transverse residual stress under different preheating temperatures. 92
4.38 The cooling rate at selected points for preheating effect. 92
4.39 Comparison of transverse residual stress under different weld lumping schemes. 93

4.40 The average cooling rate at selected points for different lumping schemes. 94
4.41 Comparison of transverse residual stress under different welding speeds. 95
4.42 The average cooling rate at selected points for welding speed effect. 95
4.43 Comparison of transverse residual stress under different welding sequences. 96
4.44 The average cooling rate at selected points for welding sequence effect. 96
5.1 The geometry of the box T-joint (all dimensions in mm). 101
5.2 Fabrication procedure of HSS box T-joint. 102
5.3 The cross section and welding sequence of the box section (all dimensions in mm). 103
5.4 The welding sequence of cross section at the intersection (all dimensions in mm). 103
5.5 Operation in the welding process for preheated box hollow section. 104
5.6 Welding direction for joint fabrication. 105
5.7 Hole geometry and residual stresses. 105
5.8 Revised drilling setup. 106
5.9 Close view of the drilling setup. 106
5.10 The strain gauges for the specimen with preheating. 107
5.11 The strain gauges for the specimen at ambient temperature. 108
5.12 Strain gauge scheme around Corner *b*. 108
5.13 Strain gauge scheme around Corner *d*. 108
5.14 Residual stress calculation procedure. 109
5.15 Sign conventions for residual stresses. 110
5.16 The maximum principal stress distribution for the specimen with preheating (position: 10/15 mm, unit: MPa). 110
5.17 The maximum principal stress distribution for the specimen with preheating (position: 20/25mm, unit: MPa) 110
5.18 The transverse residual stress distribution for the specimen with preheating (position: 10/15mm, unit: MPa). 111
5.19 The transverse residual stress distribution for the specimen with preheating (position: 20/25mm, unit: MPa). 111
5.20 The longitudinal residual stress distribution for the specimen with preheating (position: 10/15mm, unit: MPa). 112
5.21 The longitudinal residual stress distribution for the specimen with preheating (position: 20/25mm, unit: MPa). 112
5.22 The transverse and longitudinal residual stress at the chord weld toe (preheated specimen). 113
5.23 The relationship between the transverse and longitudinal residual stress with the distance from the weld toe (preheated specimen). 113
5.24 The principal stress distribution for the specimen welded at ambient temperature (position: 10/15 mm, unit: MPa). 115
5.25 The principal stress distribution for the specimen welded at ambient temperature (position: 20/25 mm, unit: MPa). 115

5.26 The transverse residual stress distribution for the specimen welded at ambient temperature (position: 10/15 mm, unit: MPa). 115
5.27 The transverse residual stress distribution for the specimen welded at ambient temperature (position: 20/25 mm, unit: MPa). 116
5.28 The longitudinal residual stress distribution for the specimen welded at ambient temperature (position: 10/15mm, unit: MPa). 116
5.29 The longitudinal residual stress distribution for the specimen welded at ambient temperature (position: 20/25mm, unit: MPa). 117
5.30 The variation of transverse and longitudinal residual stress at the chord weld toe for the specimen with preheating. 117
5.31 The relationship between the transverse and longitudinal residual stress with the distance from the weld toe. 117
5.32 Comparison between the transverse residual stresses of two specimens. 119
5.33 Comparison between the longitudinal residual stresses of two specimens. 119
5.34 Study for the chord edge effect. 120
5.35 The transverse residual stress variation along hole depth (preheated specimen). 121
5.36 The transverse residual stress variation along hole depth (preheated specimen). 121
5.37 The transverse residual stress variation along hole depth (preheated specimen). 122
5.38 The transverse residual stress variation along hole depth (preheated specimen). 122
6.1 Overall model used in the analysis. 127
6.2 Weld near the joint intersection. 127
6.3 Geometry of weld in box and joint intersection (all dimension in mm). 128
6.4 Setting for thermal intersection in box formation. 130
6.5 Welding speed in chord-brace intersection. 131
6.6 Modeling steps for the box joint. 132
6.7 Mesh for the model in heat transfer analysis. 133
6.8 Details of a corner of the box section. 133
6.9 Position and numbering of monitoring points. 133
6.10 Temperature field at selected moments. 134
6.11 Temperature fields comparison between the weld toe and at 15 mm position. 135
6.12 Cooling rate comparison between the weld toe and at 15 mm position. 135
6.13 Comparison for the transverse residual stress between modeling and testing (preheated). 136
6.14 Comparison for the transverse residual stress between modeling and testing (ambient temperature). 137
6.15 Temperature field at selected moments (ambient temperature). 138

6.16 Selected cross sections in the joint. 139
6.17 von Mises residual stress at 1–1' cross section (ambient temperature). 139
6.18 von Mises residual stress at 2–2' cross section (ambient temperature). 140
6.19 von Mises residual stress at 3–3' cross section (ambient temperature). 141
6.20 Welding start locations selected for analysis. 142
6.21 The actual welding start position and weld path direction. 143
6.22 Comparison of residual stress with different joint angles (*joint angle: 90°, starting location of welding: Point 1, ambient temperature, b/c: 0.66, welding speed: 2.8 mm/s*). 143
6.23 Comparison of residual stress with different joint angles (*joint angle: 135°, starting location of welding: Point 1, ambient temperature, b/c: 0.66, welding speed: 2.8mm/s*) 144
6.24 Comparison of residual stress with different joint angles (*starting location of welding: Point 1, ambient temperature, b/c: 0.66, welding speed: 2.8 mm/s*). 144
6.25 Comparison of residual stress with different joint angles (*starting location of welding: Point 1, preheating temperature: 100 °C, b/c: 0.66, welding speed: 2.8 mm/s*). 145
6.26 Comparison of residual stress with b/c value (*joint angle: 90°, starting location of welding: Point 1, ambient temperature, welding speed: 2.8 mm/s*). 146
6.27 Comparison of residual stress with b/c value (*joint angle: 90°, starting location of welding: Point 1, preheating temperature: 100 °C, welding speed: 2.8 mm/s*). 146
6.28 Comparison of residual stress with different welding start location (*joint angle: 90°, preheating temperature: 100 °C, b/c: 0.50, welding speed: 2.8 mm/s*). 147
6.29 Comparison of residual stress with different welding start location (*joint angle: 90°, preheating temperature: 100 °C, b/c: 0.33, welding speed: 2.8 mm/s*). 147
6.30 Comparison of residual stress with different welding speeds (*joint angle: 90°, ambient temperature, b/c: 0.67, welding starting location: Point 1*). 148
6.31 Comparison of residual stress with different welding speeds (*joint angle: 90°, preheating temperature: 100 °C, b/c: 0.67, welding starting location: Point 1*). 149
6.32 Comparison of residual stress with different welding speeds (*joint angle: 90°, preheating temperature: 100 °C, b/c: 0.50, welding starting location: Point 1*). 149
6.33 Comparison of residual stress with different welding speeds (*joint angle: 90°, preheating temperature: 100 °C, b/c: 0.33, welding starting location: Point 1*). 150
7.1 Test rig and specimen installation. 154

7.2 Actuators in three directions. 154
7.3 The dimensions of the box T-joint and ends connection (all dimension in mm). 155
7.4 The dimensions of the elongation parts for chord (all dimension in mm). 155
7.5 Plane view of strain gauge location in preheated joint. 156
7.6 Enlarged view of the strain gauge scheme. 157
7.7 Extrapolation zone near the chord weld toe. 157
7.8 Strain gauge put around the intersection. 158
7.9 A close view of strain gauge put around the intersection. 158
7.10 SCF distributions on chord box under axial loading. 160
7.11 SCF distributions on chord box under in-plane bending. 160
7.12 SCF distributions on chord box under out-of-plane bending. 161
7.13 RSF distributions at selected points under axial loading (ambient temperature specimen). 162
7.14 RSF distributions at selected points under axial loading (preheated specimen). 163
7.15 RSF distributions at selected points under in-plane loading (ambient temperature specimen). 163
7.16 RSF distributions at selected points under in-plane bending (preheated specimen). 164

List of Tables

2.1 Distribution of average critical linear elastic fracture mechanics parameter, K_{IC} (MPa.m$^{0.5}$), at room temperature (20 °C). 12
2.2 Codes and standards of HSS. 13
2.3 Comparison for testing methods of residual stress. 25
2.4 Couplings happened in the welding process. 27
3.1 Mechanical properties of RQT701 steel plate and LB-70L electrode. 40
3.2 Chemical composition of the RQT701 steel plate and the LB-70L electrode. 40
3.3 Geometry of the specimens (Note: width of all specimens = 150 mm). 41
3.4 Welding specification for the specimens. 42
3.5 Coefficients $\bar{a}$ and $\bar{b}$ at different depths. 46
3.6 Residual stress measurement results. 48
3.7 Summary of SCF values. 57
4.1 Summary of the sequentially-coupled thermal-stress analysis procedure. 66
4.2 Modeling results at selected points (total: **18** models). 71
4.3 Summary of the modeling for parametric study (total: 23 models). 88
5.1 Material properties at room and elevated temperatures. 100
5.2 Welding parameters for box sections and joints. 101
5.3 Residual stress of the preheated specimen. 113
5.4 Residual stress of the specimen welded at ambient temperature. 117
5.5 Residual stress variation along depth in corners of the specimen with preheating. 122
6.1 Comparison of the modeling procedure between plate-to-plate and box joints. 131
6.2 Range of the parametric study. 142
7.1 Geometrical parameters of the T-joints. 156
7.2 Comparison for SCF values between test results and CIDECT equations. 161

Preface

High-strength steel structures have unique characteristics and outstanding performance, with clear advantages in construction applications. However, research on high-strength steel structures is not sufficiently in-depth, which to some extent limits the application and development of these structures. Therefore, we should pay attention to the application of high-strength steel structures by understanding the key points of construction quality control and improving construction quality and efficiency.

The purpose of this book is to provide a text dealing exclusively with welding and its consequences for high-strength steel structures. It is intended that the book should cover the most important subjects related to performance of high-strength steel connection. The residual stress of welded plate-to-plate T/Y-joints and box joints is experimentally and numerically studied in this book. When high-strength steel is to be used in tubular structures, it is needed to form the box sections firstly by welding. Box hollow section joint is then fabricated by welding the profiled ends of the braces onto the surface of the chord. When the high-strength steel box joint is under cyclic loading, fatigue failure initiates at the intersection part due to the stress concentration effect which is caused by the weld profile and residual stress. Thus, in this book, I have extended the research on welding of high-strength steel structures by focusing on the residual stress and its effect on stress concentration for plate-to-plate T/Y-joints and box T/Y-joints fabricated using HSS with yield stress up to 690MPa.

Regarding the organization of this book, there are eight chapters. Chapter 1 gives research background and application of high-strength steel structures. Chapter 2 gives the summary of the research progress high-strength steel structures. Chapter 3 and Chapter 4 give experimental and numerical investigation of plate-to-plate T/Y-joints. Chapter 5 to Chapter 7 give experimental and numerical investigation of box T/Y-joints. Chapter 8 gives conclusion and recommendation.

Many friends and colleagues have contributed to this book by providing application examples and literature references, by reviewing draft chapters, and by supporting me in other ways. I want to acknowledge my dear friends and colleagues who make great contribution to this work.

Notation

The notation used in the main text is briefly defined in the alphabetical order in the following list. Symbol whose use is confined to only a specific analysis is omitted here and is defined at the appropriate point in the text.

$\overline{a}$	Calibration constant for normal stresses
$\overline{b}$	Calibration constant for shear stresses
c	Specific heat
f_y	Yield stress of steel
h	Convection coefficient
h_w	Height of a weld lump
k	Material conductivity
l	Weld filler length in the chord plate surface
l_1	The length of the chord plate
l_2	The length of the brace plate
l_w	Width of a weld lump
q	Heat flux from outside into the body
q_c	Net rate of convection heat transfer
q_r	Net rate of radiation heat transfer
r	Heat flux generated in the body
t	Heat propagation time (time after the welding started)
t_1	The thickness of the chord plate
t_2	The thickness of the brace plate
D	Gauge circle diameter
D_0	Diameter of the drilled hole
$\dot{E}_{gen}$	Heating rate generated from a weld lump
E	Young's modulus
I	Arc current
K_t	Average cooling rate
Q	Heat input
R	Weld root length

T_p	The preheating temperature
T	Temperature
T_0	The ambient temperature (30 °C)
T_b	Benchmark temperature
U	Arc voltage of welding
α	Clockwise angle from the x-axis to the maximum principal stress direction
β	End preparation angle for the specimens
ε	Relieved strain due to the hole-drilling
ϑ	Joint angle
ν	Poisson's ratio
σ_c	Calibration stress
σ_x	Uniform normal stress in x direction
σ_y	Uniform normal stress in y direction
σ_1	Maximum principal stress
σ_2	Minimum principal stress
σ_B	The Stefan-Boltzmann coefficient
τ_{xy}	Uniform shear stress
ν	Poisson's ratio
ϑ	Joint angle
δ_θ	An arbitrary variation of the temperature field
Δ_t	Time increment in the modeling
Δl	Distance between nodes
$[C]$	Heat capacitance matrix
$[K]$	Thermal conductivity matrix
$\{Q\}$	External flux vector
$\{T\}$	Temperature field
$\{\sigma\}$	Stress field
AWS	American Welding Society
CE	Equivalent carbon content
HAZ	Heat-affected zone
HSS	High-strength steel

1

Introduction

1.1 Research Background

Currently, most steel structures are made of mild steel for its satisfactory mechanical properties and availability. However, there has been an increasing interest in the use of high-strength steel (HSS), which generally has yielding strengths larger than 460 MPa, due to recognition of the benefits from an increase in the strength-to-weight ratio and savings in the cost of materials. One major application of HSS is in jack-up structures. Obvious increases of its application have occurred in the topside areas of jacket structures where the weight saving not only produces overall saving in materials but also allows crane barge installation of more complete topside processing and accommodation units with significant savings. Some HSSs with yield strengths up to 700 MPa have been used in mobile jack-up drilling rigs to minimize weight during the transportation stage (Billingham et al. 2003). In some offshore structures, such as the BP Harding Jack-up and Siri field production Jack-up, HSS has been successfully used.

Another application of HSS is in building and bridge construction. Application of HT780 high-strength steel plate to structural member of super high-rise building can be found (Mochuzihi et al. 1995; Hagiwara et al. 1995).The roof truss of the Sony Center in Berlin sets an example for the use of high-strength steel S690 in buildings (Sedlacek and Muller 2005). This steel S690 is aimed to protect an old masonry building by suspending several stories of the building. The truss structure is made of steel S460 and S690 and it is formed in solid rectangular shapes to keep the dimensions of the cross section small (Figures 1.1 and 1.2). Some examples of the application of HSS can also be found in Asian countries. For example, in Japan Landmark Tower in central Yokohama is the first project using HSS in the construction of buildings. Another application of HSS with minimum tensile strength 600 MPa is the Shimizu super high-rise building, which is 550 m high and comprised of 127 levels, where HSS was used for reducing the column section size. HSS research in Australia was first launched by Rosier and Croll and it was used in the project of Grosvenor Place in Sydney in 1989.

It can be found that HSS is more widely used in construction engineering for its advantages in economy, architecture, sustainability, and safety when compared

Welded High Strength Steel Structures: Welding Effects and Fatigue Performance, First Edition. Jin Jiang.

Figure 1.1 Sony center in Berlin.

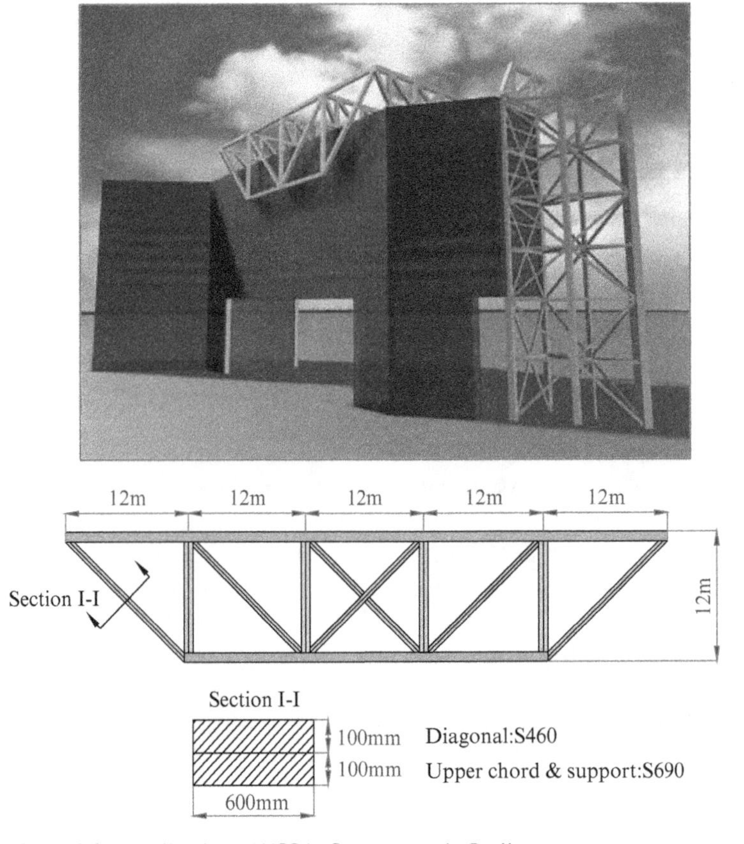

Figure 1.2 Application of HSS in Sony center in Berlin.

with mild steel. The weight and size of the structural sections can be reduced by using HSS and hence fabrication and erection costs can be reduced. Aesthetic and elegant structures can be achieved by using HSS to reduce the size of structural sections. In addition, the use of HSS means reduced consumption of raw materials and is beneficial in sustainability. Figure 1.3 shows the development of the production processes for rolled steel products. It is shown in Figure 1.4 that by increasing the strength of steel the structural section can be reduced, followed by a reduction of structure weight (Sedlacek and Muller 2005).

On the other hand, the stress-strain behavior of HSS is different from the stress-strain behavior of mild steel, since HSS shows reduced capacity for strain hardening after yielding as well as reduced elongation. Therefore, HSS generally has lower ductility than mild steel. Also, HSS is difficult to weld due to its much higher carbon and alloy contents than mild steel. The residual stress due to the welding may have

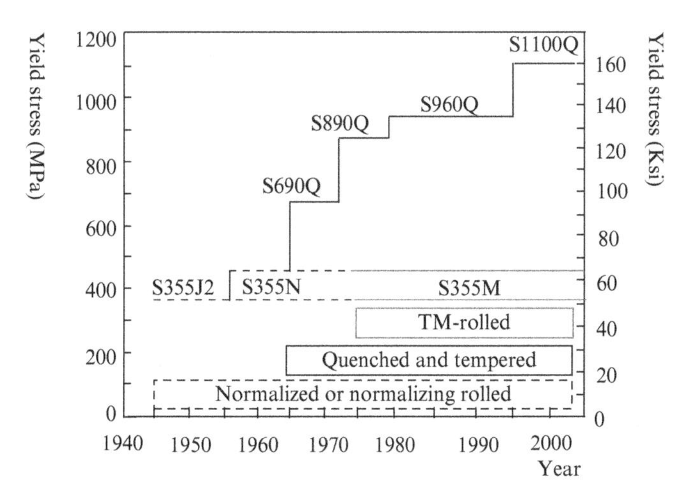

Figure 1.3 Historical development of production processes for rolled steel products (Sedlacek and Muller 2005).

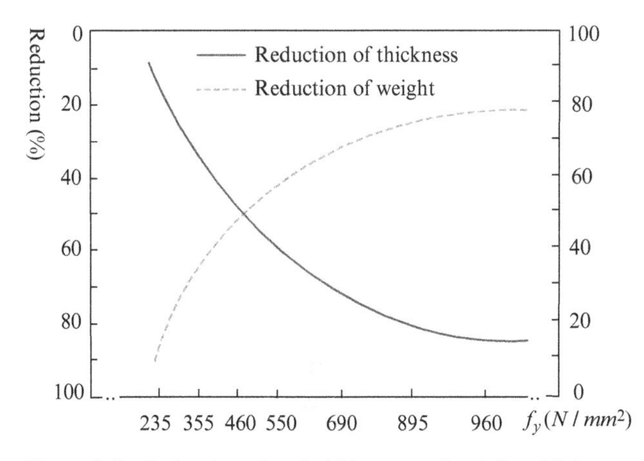

Figure 1.4 Reduction of wall thickness and weight with increasing strength of steel (Sedlacek and Muller 2005).

significant influence on the fatigue performance of HSS structures. In addition, fatigue is one of the major problems causing the HSS structures' degradation in long term integrity when cyclic load exists.

Unlike mild steel, which can be provided in rectangular hollow section (RHS) and circular hollow section (CHS), HSS, especially for the steels with yielding stress larger than 690 MPa, is mainly supplied in plate form. Therefore, when HSS is to be used in tubular structures, it is necessary to form the box sections firstly by welding. Box hollow section joints are then fabricated by welding the profiled ends of the braces onto the surface of the chord. When the HSS box joint is under cyclic loading, fatigue failure initiates at the intersection part due to the stress concentration effect, which is caused by the weld profile and residual stress. However, most fatigue design standards only cover the steels with the yielding stress less than 500 MPa. Fatigue performance data for the welded HSS box section joint is rare. In addition, many investigations of the fatigue analysis for RHS and CHS joints assume that the residual stress due to the welding in the intersection of brace and chord can be ignored. However, this assumption may not be suitable for HSS box section joints. The distribution of welding residual stress in HSS box joints and the influence of welding parameters on the residual stress files are still lacking data.

Welding method, welding position, and welding parameters such as welding speed, preheating temperature, and weld pass are of importance in view of residual stress. The welding residual stress field can be diverse when different welding processes are used. In practice, arc welding prevails in steel structure construction because of its simplicity and ease of handling in the field. Welding power is supplied to create and maintain an electric arc between the base material and an electrode to melt metals at the welding position. Due to highly localized heat input during the arc welding, microstructure and properties in the base material near the weld are altered. The cooling rate at different locations of the base material is different, causing thermal strain and stress in the welding process. In order to investigate the welding residual stress on the welded joints, many experimental works can be found (Teng et al. 2001). However, most of these works are based on simple butt welds with base material less than 460 MPa. In addition, numerical modeling of residual stress has advantages in understanding the impact of welding parameters. However, few works on the modeling of the residual stress in HSS box T/Y-joints can be found in the literature.

1.2 Objectives and Scope

Based on the discussion above, the main objectives of this book focus on the residual stress and its effect on stress concentration for plate-to-plate T/Y-joints and box T/Y-joints fabricated using HSS with yield stress up to 690 MPa. In particular, attentions are given in the following areas:

- To experimentally investigate the residual stress distribution of welded plate-to-plate T/Y-joints which are fabricated with HSS with yielding stress up to 690 MPa.
- To numerically model the residual stress fields of the HSS plate-to-plate T/Y-joints corresponding to the specimens used in experimental investigation and to validate the accuracy of the numerical models.

- To experimentally investigate the residual stress distribution of welded HSS box T-joints and to evaluate the effect of preheating on the residual stress magnitude near the intersection of the joints.
- To experimentally and numerically investigate the stress concentration factors in the intersection part of the HSS box joints under basic and combined load cases.
- To numerically model the residual stress fields of the HSS box T/Y-joints corresponding to the specimens used in experimental investigation and to organize parametric study to evaluate the impact of joint geometry and welding parameters on the residual stress field.

In order to achieve the proposed objectives, the scope of this book is outlined below. At the same time, Figure 1.5 shows the organization of the thesis.

- To explore the residual stresses distribution in welded HSS plate-to-plate joints. To find the distribution of residual stresses at different conditions, 18 plate-to-plate HSS joints, with different intersection angles, base plate thicknesses, preheating temperature, and cutting treatment, will be examined.

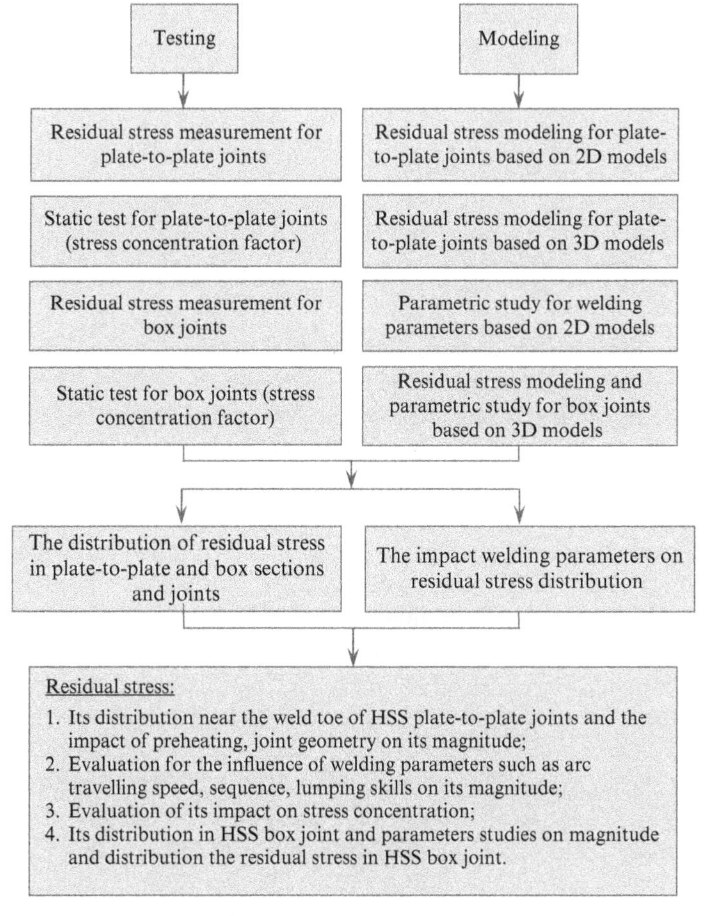

Figure 1.5 Technical flow of the book.

- To simulate the residual stress fields for the joints corresponding to the specimens used in experimental investigation based on 2D and 3D modeling. In addition, by conducting a small-scale parametrical study, the impacts of welding parameters such as arc travelling speed, welding sequence, and preheating temperature on the distribution of residual stress will be examined.
- To experimentally study the residual stress due to the weld in the chord and brace box and to investigate the residual stress distribution around the chord weld toe when the HSS box joint is formed. Two specimens, one welded at ambient temperature and the other preheated to 100 °C before welding, will be tested to evaluate the effect of preheating on the residual stress magnitude. To investigate the stress concentration factor (SCF) distributions of welded box T-joints under the basic and combined load cases experimentally and numerically. Three basic loads, axial load, in-plane bending, and out-of-plane bending, and their combinations will be applied to check the stress concentration effect due to the weld profile and fabrication procedure near the chord and brace intersection part of the welded box T-joints.
- To numerically model the residual stress fields of the HSS box chord and brace formed by welding and to show the final residual stress fields when the welded HSS box T-joints are formed. A parametric study is conducted to investigate the impact of geometry of joint and welding parameters based on 3D models.

1.3 Contributions and Originality

The main contributions of this book are to provide the residual stress fields for HSS plate-to-plate joints and box joints from experiments and numerical analyses. The main contributions and originalities of this study are outlined below:

- Experimental investigation of welding residual stress of HSS plate-to-plate joints with different geometries and welding treatments and evaluation of the influence of joint geometries and preheating temperature on the residual stress field. The originality of this part includes revision for the HS-200 drilling setup to fit the specimens used in the testing and finding of the preheating effect of the residual stress for HSS plate-to-plate joints.
- Numerical investigation of welding residual stress of HSS plate-to-plate joints based on 2D and 3D models and evaluation of the impact of welding parameters such as preheating temperature and welding speed on the residual stress magnitude. The originality of this part includes finding for the effect of welding parameters (preheating temperature, welding speed, welding sequence, welding direction), mechanical boundary condition, and modeling skill (lumping way) on the residual stress field.
- Experimental investigation of welding residual stress of HSS box chord, brace sections and T-joints and experimental study of the stress concentration factor of HSS box joints under basic and combined loads. The originality of this part

includes finding for the residual stress distribution in HSS box joints and effect of the welding residual stress on the stress concentration.
- Numerical investigation of welding residual stress of HSS box chord, brace sections, and T- and Y-joints based on 3D models. The originality of this part includes the application for fully coupled thermo-mechanical analysis in welding residual stress simulation. The evaluation for the influence of welding parameters, including welding speed, preheating temperature, and joint geometry on the residual stress is another main originality in this part.

1.4 Organization

The organization of this study is arranged as follows:

- Chapter 2 covers literature review. The relevant research and work will be reviewed and the background of related theories will be introduced.
- Chapter 3 introduces the experimental investigation of residual stress of HSS plate-to-plate joints. The test method, specimens, test procedure, and findings will be introduced. The residual stress near the weld toe of the HSS plate-to-plate joints will be shown. Eighteen joints with different joint angles, plate thicknesses, and preheating treatments will be included to study the residual stress.
- Chapter 4 investigates 2D and 3D numerical modeling for residual stress of HSS plate-to-plate joints. The modeling techniques used will be covered. The modeling results will be validated with the testing results from Chapter 3. A small-scale parametric study will be carried out after the modeling validation to investigate the impact of welding parameters on the residual stress field.
- Chapter 5 gives the experimental investigation of residual stress of HSS box T-joints. In this chapter, the residual stress field around the chord weld toe of the HSS box T-joints will be shown. Comparison of residual stress between two box T-joints, one welded at ambient temperature and the other welded with preheating, will be organized to evaluate the effect of preheating.
- Chapter 6 shows 3D numerical modeling for the residual stress of HSS box T-joints. The geometrical modeling for the joints will be introduced first in this chapter. The residual stress fields for both joints will be shown next. A parametric study will be carried out to investigate the influence of preheating temperature, welding start position, welding speed, and joint angle on the residual stress field.
- Chapter 7 covers the full-scale static test of two box T-joints for investigating the stress concentration factors around the chord weld toe under basic and combined loads. The test rig and test procedure will be introduced and the SCFs around the weld toe in different loads cases will be shown.
- Chapter 8 summarizes the results and observations in this thesis. Conclusions are drawn based on the experimental and numerical works and some recommendations will also be given in this chapter for future works.

2

Literature Review

2.1 High-Strength Steel

2.1.1 Overview

Generally, steel strength is determined by the steel microstructure which is related to the steel chemical composition, thermal history, and deformation process. One way to produce HSS is through quenching and tempering. With this technique, HSS can be made by rapidly cooling an austenitized specimen in a quenching medium to form martensite microstructure and balancing steel ductility by controlled heating the quenched work piece. To increase the tensile strength of the material, the HSS steel is heated above the upper critical temperature and then cooled quickly. Upon being cooled rapidly, a portion of austenite (dependent on alloy composition) will transform to martensite, which is a hard and brittle crystalline structure. After that, tempering is applied for HSS by heating steel below the lower critical temperature to improve the toughness of the steel. In practice, appropriate tempering temperatures are selected to produce the required level of hardness and strength.

Compared with mild steel, HSS shows reduced capacity for strain hardening after yielding since the steel strengthening mechanism is used to increase the yield strength. Therefore, yield ratio, defined as the ratio of yield strength to ultimate strength of HSS, is lower than that of mild steel. There is a restriction in structural design that states that yield ratio is not allowed to have a value greater than 0.85 in design equations in order to ensure an adequate ductility in the structural member to develop plastic failure behavior as a defense against brittle fracture (Billingham et al. 2003). However, it is necessary to note that the yield ratio is not directly related to the capability of a given steel to withstand plastic strain after yielding and before fracture. Elongation generally decreases as yield ratio increases for older HSS, while it still has significant elongation for grade 690 with yield ratio of 0.95 when it is made of modern clean steels with low carbon content and low levels of impurity.

Welded High Strength Steel Structures: Welding Effects and Fatigue Performance, First Edition. Jin Jiang.

2.1.2 Delivery Condition of HSS

The as-rolled condition of steel includes the rolling of steel slab at the austenitic state after heating at temperatures of about 1100 °C and the cooling process to room temperature on air (Easterling 1992). The normalized condition (N) is to reheat the cooled-down steel slab above the ferrite-austenite transformation temperature (about 800–900 °C) to get a more homogenous microstructure (Easterling 1992). The transformation of ferrite and pearlite to austenite and back again is achieved in this period. This leads to a refined microstructure of ferrite and pearlite. Steels with moderate strength and toughness requirements up to S460N can be produced by normalizing.

The objective of quenching and tempering (QT) procedure is to produce a metal microstructure consisting mainly of tempered martensite. Quenching is a metallurgical processing in which the material is rapidly cooled down to obtain high hardness (Bozidar 2010). Quenching of HSS is frequently performed after austenitizing at temperatures of 900 °C. When quenching is performed, the material can be rapidly cooled with air, liquid polymers, oil, or water. For the different quenching methods and the metal thicknesses, the cooling rate of the material can be diverse. The hardenability requirement may be satisfied after the quenching treatment. However, the quenched material is too brittle for most structural applications when considering the mechanical properties. Therefore, tempering is performed after the quenching to transform brittle martensite or bainite into a combination of ferrite and cementite by reheating of the work piece to a temperature below its lower critical temperature. Therefore, there is always a trade-off between strength and ductil ity for the HSS steel. Figure 2.1 shows the comparison of temperature history for different steel delivery conditions and Figure 2.2 gives an example for the microstructures of steel when different delivery conditions are applied.

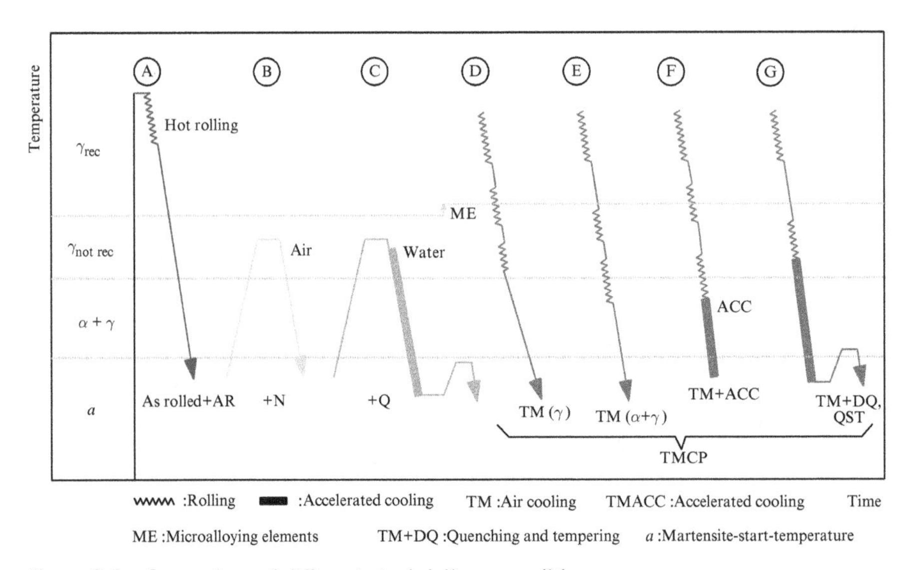

Figure 2.1 Comparison of different steel delivery conditions.

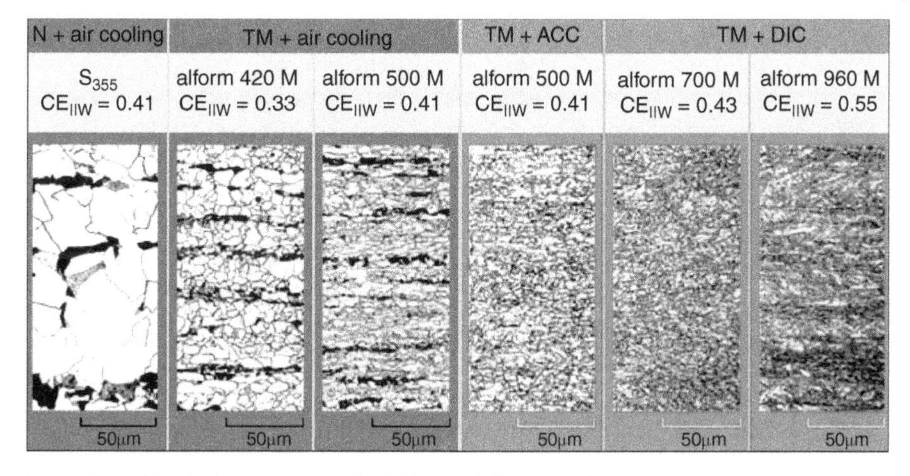

Figure 2.2 Steel microstructures for different delivery conditions.

Another method to get HSS with an extremely fine-grain microstructure is thermo-mechanical control process (TMCP) (Willms 2009). There exists an inverse relationship between the grain size and the tensile and toughness properties. TMCP creates HSS by controlling conditions during the steel plate manufacturing process to get much finer microstructure compared with conventional steel. It is achieved by a skilled combination of rolling steps at particular temperatures and a close temperature control (Granjon 1991). Compared with normalized steel, the carbon and alloying content of TMCP steel is lower when the same steel grade is considered. Figure 2.3 gives the steel strength under different delivery conditions.

2.1.3 Fatigue and Fracture of HSS

Fatigue is one of the major problems causing the degradation of offshore structures in long term integrity. The loadings applied to structures or components are not

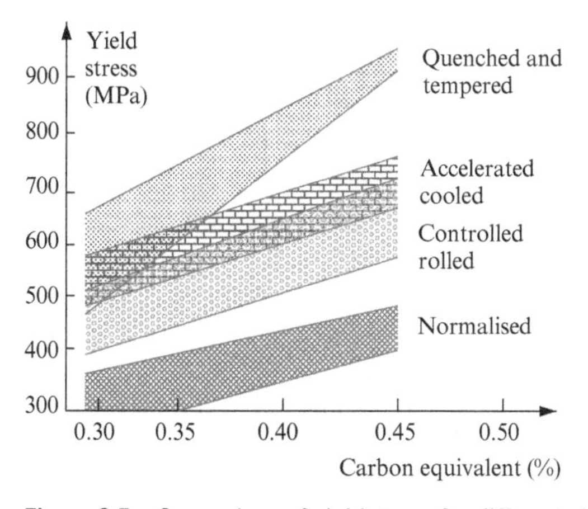

Figure 2.3 Comparison of yield stress for different delivery conditions.

necessarily large enough to cause immediate failure and it may occur after a certain number of loading fluctuations have been experienced. It is found that a great proportion of total repair works are induced by fatigue damage in offshore steel jack-ups, and consequently the fatigue is considered as one of the most significant failure models in the evaluation offshore steel structures (NORSOK STANDARD 2004).

Many researchers have shown their interests in fatigue performance of HSS since the 1990s (Agerskov and Petersen 1998; Clarin 2004; Healy and Billingham 1995; Petersen et al. 1996; Prior and Maurer 1995; Sharp et al. 1999; Stacey et al. 1996). Chen (Chen 2005; Chen et al. 2003) proposed that the parent steel grade does not have obvious influence on fatigue life. However, this conclusion was obtained based on specimens with special heat treatment for welding or intact specimens without welding.

It was commented (Maddox 1991) that the use of HSS may be of advantages when the expected number of nominal load cycles is comparatively small or when the mean stress is high so that the fatigue strength is not the dominating design factor. Anami and Miki mentioned that the magnitude of residual stress increases with the material yield strength, and a higher residual stress in welded HSS connections could lead to a reduction in fatigue strength when compared with welded mild steel connections. Anami and Miki (Anami and Miki 2001; Miki et al. 2002) reviewed the development and use of HSS for bridge structures in Japan. It was found that the most important issue for the application of HSS is to achieve a balance between tensile strength and fatigue performance without losing good weldability.

Fracture toughness is a property which describes the ability of a material containing a crack to resist fracture. Fracture toughness is an indication of the amount of stress required to propagate a preexisting flaw. A previous study (Hajro et al. 2010) analyzed the fracture toughness of some high-strength steels (S690 and S890), as listed in Table 2.1. However, for S460, no data is found from published work.

2.1.4 Codes and Standards of HSS Application

The application of HSS in construction engineering is limited by codes and standards since the design formulae in existing standards are restricted to steels with yield strength smaller than 500 MPa in most cases. However, some existing standards and codes can be found related to HSS. Materials specifications, welding requirements, and mechanical properties are partially included. Table 2.2 records the existing standards and codes on HSS in detail.

Table 2.1 Distribution of average critical linear elastic fracture mechanics parameter, K_{IC} (MPa.m$^{0.5}$), at room temperature (20 °C).

Category	Base metal (BM)	Heat-affected zone (HAZ)	Coarse-grained heat-affected zone (CGHAZ)	Weld
S690	227	205	137	168
S890	202	177	107	166

Table 2.2 Codes and standards of HSS.

Item	Code, Standard/ Report	Scope and comments
Materials specifications	ASTM A514 (ASTM 2005)	Quenched and tempered steel plates of structural quality in thicknesses of 150 mm and under intended primarily for use in welded bridges and other structures.
	DnV OS-B101 (DNV 2008)	Steel grade is covered to 690 MPa.
Welding specification	AWS D1.1:2008 (AWS 2008)	Covers steel grades up to 690 MPa. Welding details for different geometry types are included.
	BS EN 1011–2 2001 (BSI 2011)	Only suitable for arc welding of ferrite steels.
	AS/NZS 1554.4 Part4:2004 (AS/NZS 2004)	Limited to quenched and tempered steel. Covers steel grades up to 800 MPa.
Welding impact on HSS structure	HSE OTO 95 952 (Billingham and Spurrier 1995; HSE 1999)	The influence of welding on the HSS performance for offshore structures.
Fatigue and fracture	DnV RP-C203:2010 (DNV 2008a;DNV2008b)	Valid for steel materials in air with yield strength less than 960 MPa. For steel materials in the seawater, the Recommended Practice is valid up to 550 MPa.
	IACS (IACS 1999)	International Association of Classification Societies document. The minimum Charpy values are YS/10.
	ISO19902 (ISO 2008)	Equations provided to calculate fatigue life for the yield stress of steel larger than 500 MPa.
High temperature performance	Limited	Lack of data for HSS.
Performance of HSS tubular structure	HSE Offshore Report 2000/078 (HSE 1999)	Static strength of cracked high-strength steel tubular joints. The yielding stress of the material is 700 MPa.
	HSE OTH562 [28]	Focused on the use of high-strength steels in welded tubular joints of offshore structures.

2.2 Welding and Residual Stress

2.2.1 Overview of Arc Welding

Welding is a materials-joining process which produces coalescence of materials. In view of metallurgy, welding processes can be divided into two main categories: fusion welding processes and solid-phase welding processes. The nature of welding is to form the constitute atoms of metal, belonging initially to two separate parts (A and B), into a single assembly part, as shown in Figure 2.4 (Granjon 1991). The metal continuity is achieved by solid phase diffusion across the interface.

In fusion welding, three metallurgical zones (fusion zone, HAZ, and unaffected parent plate) are included. In solid-phase welding, two clean, solid metal surfaces are brought into sufficiently close contact with a metallic bond. The three most important characteristics of a fusion welding process are the intensity of the heat source, the heat input rate per unit length of weld, and the effectiveness of the method used to shield the weld from the atmosphere. A minimum heat source intensity is required to make the fusion weld. The heat input rate is of great importance, since it governs heating rates, cooling rates, and weld pool size. A high heat input rate generally can give less HAZ hydrogen-induced cracks in the welding. Another feature directly affected by the heat input rate is the grain size in the fusion and HAZ zone. A high heat input rate produces longer thermal cycles and tends to generate a coarser grain structure, which deteriorates the mechanical properties such as notch ductility for ferrite steels. It is therefore necessary to seek a heat input rate that gives the optimum combination of grain size and cooling rate.

Arc welding is one type of fusion welding for joining metals. Shielded metal arc welding (SMAW) and flux-cored arc welding (FCAW) are the two most common fusion welding methods. By applying intense heat, metal at the joint between two parts is melted and caused to intermix with an intermediate molten filler metal. In arc welding, the intense heat needed to melt metal is produced by an electric arc and it melts the base material and filler metal in the vicinity of the arc. As shown in Figure 2.5, the heat source is provided by a high current and a low voltage discharge in the range of 10–2000A and 10–50V. The nature of the arc is an electric current flowing between two electrodes through an ionized column of gas. During the welding, the flux in the electrode, which is composed of various silicates and metal oxides, melts to form a viscous slag that acts as a protective layer between the atmosphere and the molten metal. When metal is heated to high temperatures, it is chemically reactive with the main constituents of air, oxygen, and nitrogen. The strength of the weld joint would deteriorate when the metal in the molten pool is in contact

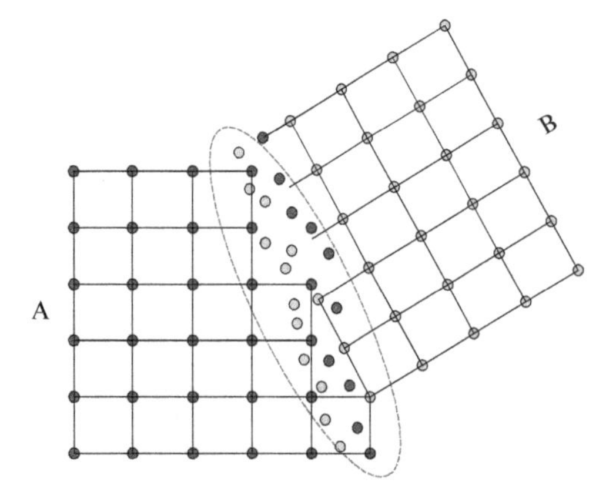

Figure 2.4 The nature of welding.

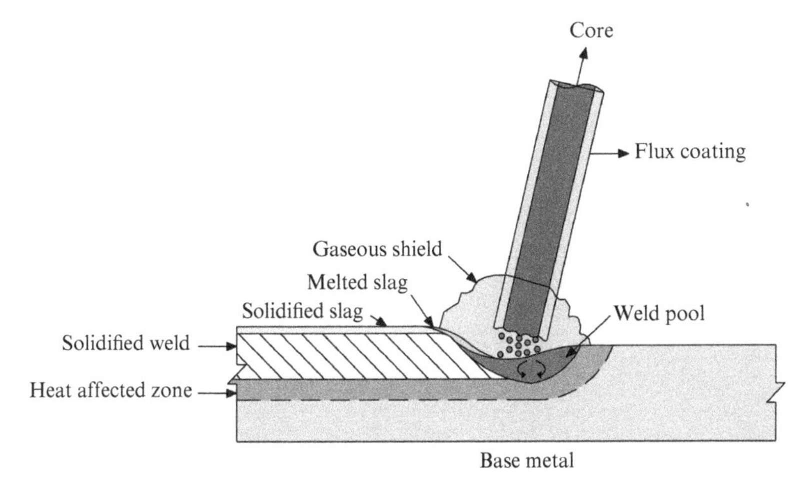

Figure 2.5 Mechanism of fusion welding.

with air, oxides, and nitrides. Therefore, shielding, such as using a vapor-generating covering over the electrode and covering the arc and molten pool with a separately applied inert gas or a granular flux, is frequently used in the construction industry.

2.2.2 Weldability of Steel

The weldability is usually expressed in terms of a carbon-equivalent limit (C_{equiv}). Weldability is defined by AWS (AWS 2008) as the capability of a metal to be welded into a specific, suitably designed structure and to perform satisfactorily in the intended service. When the C_{equiv} is less than 0.4, the steel generally is considered weldable. This value is frequently used to evaluate the effect of alloying elements on the transformation characteristics of the steel. A C_{equiv} of 0.4 is claimed by Easterling (1992) to be the maximum allowable value if defects are to be avoided. Figure 2.6 shows Easterling's findings of yielding stress as a function of C_{equiv} for a number of different steels.

Weldability varies with the chemical composition, metallurgical properties, and mechanical properties. Some of the commonly specified elements on weldability and other characteristics of steel are discussed in the following paragraphs:

Carbon is the principal hardening element in steel. There exists a positive relationship between the carbon content and tensile strength while an inverse proportion relationship can be observed between the carbon content and ductility, as well as weldability. Rapid cooling rate can produce a brittle HAZ zone when the carbon content is over 0.25% (Granjon 1991). Manganese can increase the hardenability and tensile strength of steel, but to a lesser extent than carbon. Manganese content of less than 0.30% may introduce internal porosity and cracking in the weld bead. When it is over 0.80%, cracking is also possibly produced (Granjon 1991). Manganese can increase the rate of carbon penetration during carburizing and is beneficial to the surface finish of carbon steel. Silicon increases the strength and hardness of steel, but to a lesser extent than manganese. It is detrimental to surface quality, especially for low-carbon steel. For the best welding conditions, silicon content should not

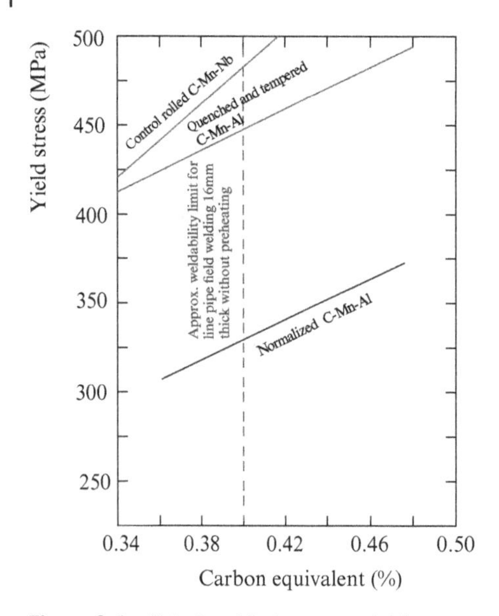

Figure 2.6 Relationship between yield stress and $C_{equiv.}$

exceed 0.10%. The influence of silicon content on weldability is not as serious as high sulfur and phosphorus content when the value of silicon content does not exceed 0.30%. Sulfur and phosphorus are harmful to impact toughness and weldability. They are also detrimental to surface quality for low-carbon steel. Generally, sulfur content can be tolerated up to 0.035%. Phosphorus should be kept as low as possible. Over 0.04% makes welds brittle and increases the tendency of cracking.

In welding, equivalent carbon content (CE) is frequently used to quantitatively describe the weldability. The two most commonly specified carbon equivalent equations are recommended by the International Institute of Welding (Eq. 2.1), which covers a wide range of steels, and the Dearden equivalent (Eq. 2.2) (Lancaster 1997), preferred for commonly-used plain carbon and carbon-manganese steels.

$$CE_{\mathrm{IIW}} = \mathrm{C} + \frac{\mathrm{Mn+Si}}{6} + \frac{\mathrm{Cr+Mo+V}}{5} + \frac{\mathrm{Ni+Cu}}{15} \tag{2.1}$$

$$CE_{\mathrm{IB}} = \mathrm{C} + \frac{\mathrm{Si}}{30} + \frac{\mathrm{Mn+Cu+Cr}}{5} + \frac{\mathrm{Ni}}{60} + \frac{\mathrm{Mo}}{15} + \frac{\mathrm{V}}{10} + 5\mathrm{B} \tag{2.2}$$

2.2.3 Phase Transformation and Other Phenomena in Welding Procedures

The phase transformation depends on the maximum temperature reached and the cooling time $t_{8/5}$ from 800 to 500 °C and steel chemical composition (Easterling 1992). This was shown by Easterling for the different metallurgical zones formed as a function of the local peak temperature in carbon steel. In Figure 2.7, three main zones can be distinguished:

The Base Metal (BM): the area where the temperature does not reach 600 °C and the microstructure is not affected by the welding operation. The steel is composed of ferrite and pearlite (a mix of ferrite and cementite).

The Heat-Affected Zone (HAZ): the area of base material with microstructure and properties altered by welding. This zone is not melted, but is transformed into austenite at high temperatures. It can be subdivided into 4 zones based on the austenite grain size:

- The Sub-Critical HAZ (SCHAZ): the area with a maximum temperature between 600 °C and Ac1 (600 °C < T_p <Ac1). Ac1 is the arrest point of temperature during heating corresponding to the beginning of the austenitization. In this area, the steel is just tempered.
- The Inter-Critical HAZ (ICHAZ): the area with a maximum temperature between Ac1 and Ac3 (Ac1< T_p <Ac3). Ac3 is the arrest point of temperature to complete austenitization. Therefore, in this area, the steel is partially austenitized.
- The Fine-Grained HAZ (FGHAZ): the area with a maximum temperature between Ac3 and 1100(Ac3< T_p <1100 °C). In this area, the steel microstructure is entirely composed of austenite grains and the steel is fully recrystallized.
- The Coarse-Grained HAZ (CGHAZ): the area with a maximum temperature exceeding 1100 °C (T_p >1100 °C). In this area, the steel microstructure is entirely composed of austenite grains and the steel is fully recrystallized. The austenite grain size is increased when compared with the FGHAZ area.

Figure 2.8 gives CCT (continuous cooling transformation) diagrams of a kind of HSS with yield stress of 500 MPa. According to the figure, for this kind of HSS, when

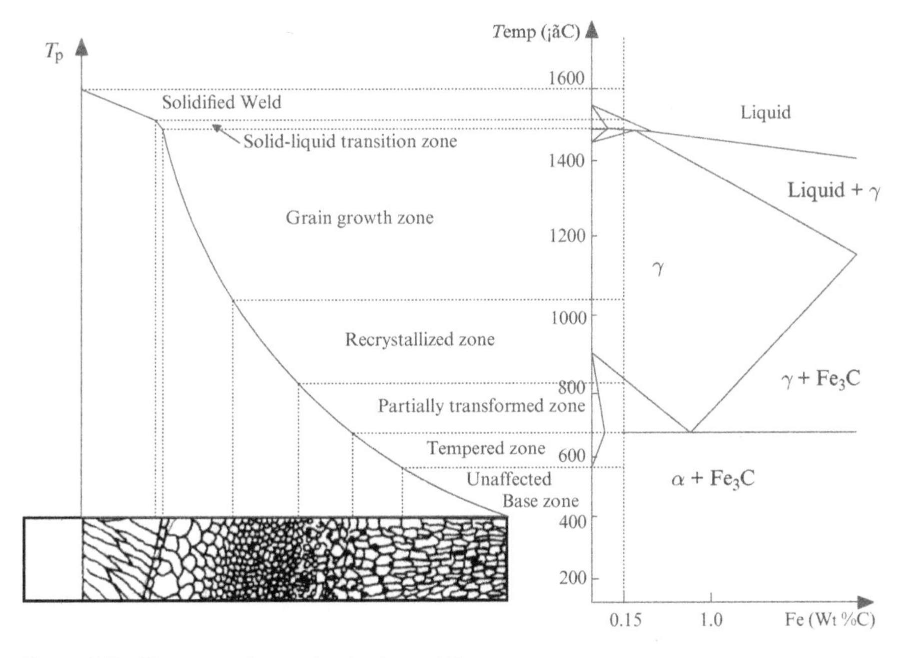

Figure 2.7 Phase transformation in the welding.

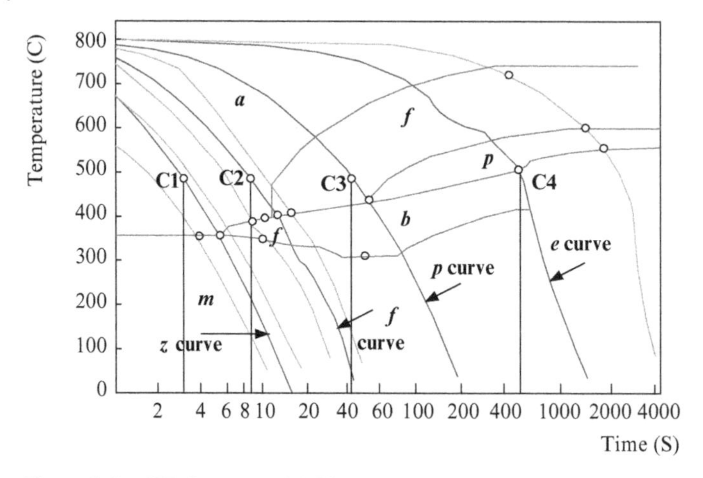

Figure 2.8 CCT diagrams of HSS.

the cooling rate is faster than z-curve, the microstructure after cooling is totally martensite. When the cooling rate becomes slow (like e-curve), the content of martensite is reduced, and the hardness of the steel will be decreased. When cooling rate is quick enough, the austenite is straightly transformed into martensite. Bainitic is formed when an intermediate cooling rate is followed. For slow cooling rates, ferrite first nucleates at the grain boundary. The non-transformed austenite is enriched by carbon and decomposes to form pearlite.

In view of metallurgy, grain size is always changed in the HAZ during the welding. It is seen as a key parameter in determining the strength and toughness of a material and determining susceptibility to cold cracking in the welds. For HSS, it is mainly based on developing a product in which the grain size is as small as possible. Therefore, increase of grain size in the HAZ will be detrimental to the strength of the material. The evolution of the microstructure of HSS is complicated since it is dependent on both temperature and deformation during the welding process. The microstructure of the HAZ is an important value since the HAZ area is the most sensitive area of the welded joint. At a macro view, it has a significant impact on the mechanical properties of the joint after welding.

The effect of welding parameters on grain structure has been studied (Kou 1988). It is described that high welding speed can elongate the weld pool, whereas the weld pool tends to be elliptical when low welding speed is applied (Figure 2.9). This phenomenon

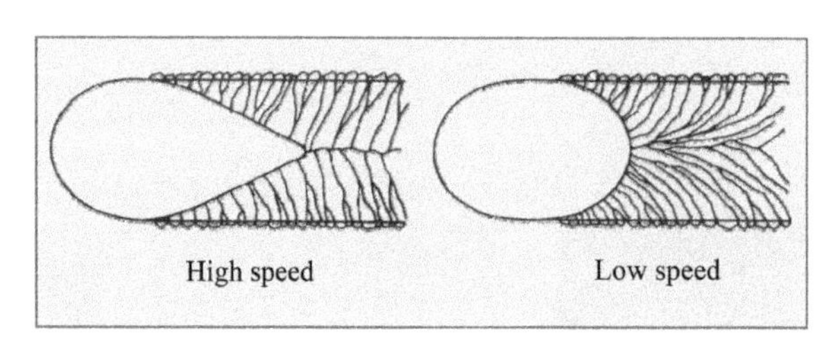

Figure 2.9 Columnar grain structure of weld.

is produced due to the boundary of the trailing portion of the tear-drop-shaped weld pool and grains are essentially straight. At the same time, the grain growing is essentially perpendicular to the weld pool boundary. A conclusion is drawn that the columnar grain will grow straight toward the weld center line when high welding speed is applied, whereas it will curve and grow in the direction of the maximum temperature gradient when low welding speed is applied.

In the welding process, hot cracks are possibly produced under high temperature and their generation is dependent on some parameters, such as plasticity temperature range, ductility, and deformation rate of the steel in plasticity temperature range. In view of mechanics, the engendering of hot cracks is ascribed to the tensile strain being larger than the critical strain of the steel (e.g. $\varepsilon > p_{\min}$ in Figure 2.10). The total strain ε and plastic strain p are functions of temperature and the curve $p = \varphi$ (T) shows the variation of plastic strain with temperature. When the strain due to the tensile stress during the crystallization goes along Line 1, $\Delta\varepsilon$ is engendered when the temperature goes to T_s, in which a hot crack has the great possibility to generate. In this case, the weld filler still has a plasticity reserve with magnitude of $\Delta\varepsilon_s = p_{\min}$-$\Delta\varepsilon$, and therefore the hot crack will not be engendered. For another case, when Line 3 is followed for strain-temperature relationship, the plastic strain is beyond the critical strain when T_s is attained, and a hot crack is generated. Cold cracks, namely hydrogen-induced cracks, are particularly associated with notches and microstructure inhomogeneity such as slag inclusions and so on. They are generated as a result of hydrogen, residual stress, microstructure, and their interactions. Figure 2.11 lists the impact parameters on the mechanical properties of welded steel structures (Radaj 1992).

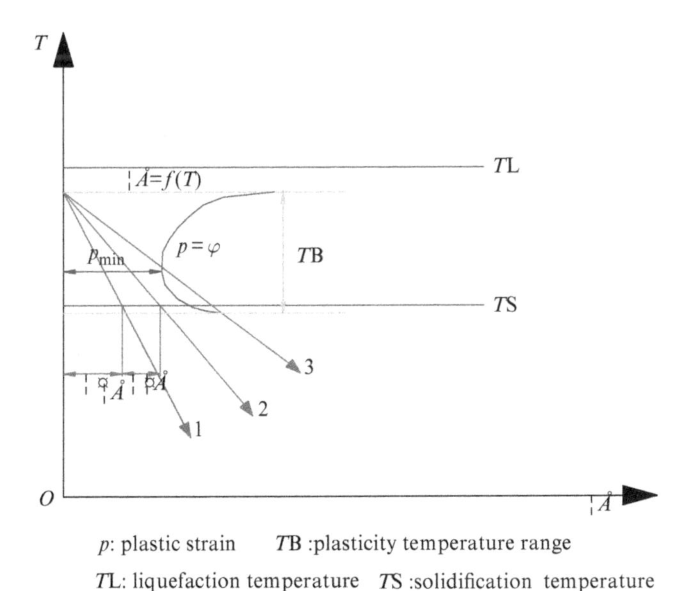

Figure 2.10 Criterion of hot crack in the weld.

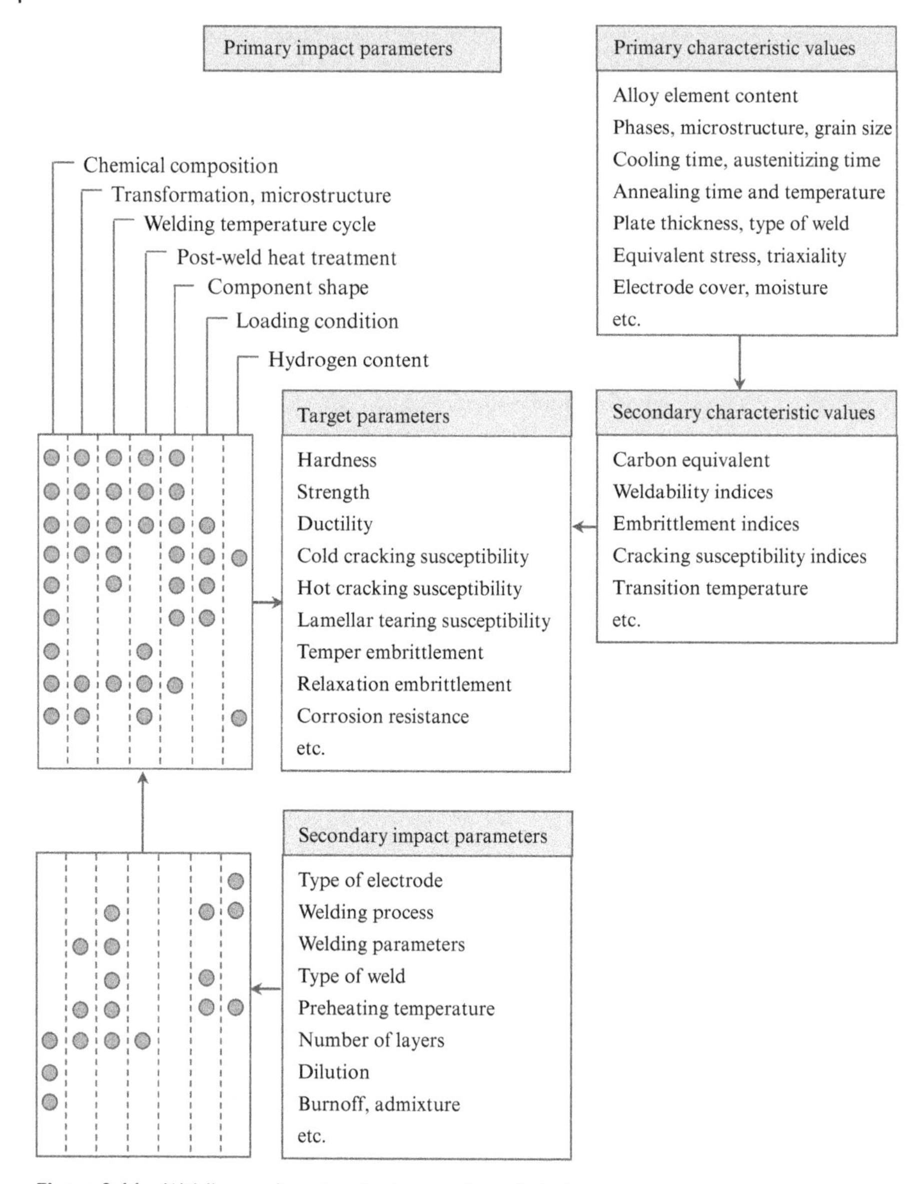

Figure 2.11 Welding and mechanical properties of steel.

2.2.4 The Formation of Residual Stress

2.2.4.1 Origin and Types of Residual Stress

Residual stress can be defined as those stresses that remain in a material or body after manufacture and processing in the absence of external forces or thermal gradients (Masubuchi 1980). The origins of residual stresses in a component can be classified into mechanical, thermal, and chemical. Residual stresses from manufacturing processes producing non-uniform plastic deformation can be assigned to mechanically generated. Operations such as welding, machining, and grinding generally produce undesirable surface tensile residual stress. Thermally generated residual stresses on a macroscopic level are often the consequence of non-uniform heating or cooling. Thermally induced stresses, which arise from temperature gradients in a material or a temperature change, have three main categories:

1) Stress induced by volumetric change, either expansion or shrinkage, associated with phase change in a material. The volume in the area near the arc torch expands in the process of heating-up during the welding process, and the extrusion effect on the surrounding material produces the inner stress in this period. When steel solidifies after being molten, its specific volume (the volume per unit mass) decreases and this volumetric shrinkage produces thermally induced stresses in the surrounding material.
2) Stress induced by a difference in the coefficient of thermal expansion (CTE) between two materials linked together, known as CTE mismatch (Beaney and Procter). When two materials with different CTEs, different expansion rates produce distortion in the materials and locked-in stress is kept in the materials. In the welding process, this will happen when two materials with different CTEs are to be welded.
3) Stress induced by temperature gradient due to different expansion rate.

This heating or cooling process can lead to severe thermal gradients which produce large internal stresses. At a selected welding moment, the residual stress at different distances from the weld pool is significantly different, as shown in Figure 2.12. Due to the different thermal cycles, the residual stress varies along the heat's moving path. In the welding process, residual stress is caused by inhomogeneous volume changes as a result of thermal expansion and the welding-induced residual stress can be described as following (Radaj 1992):

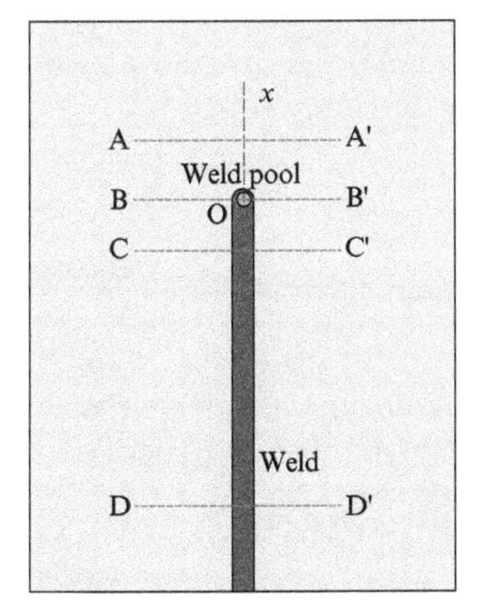

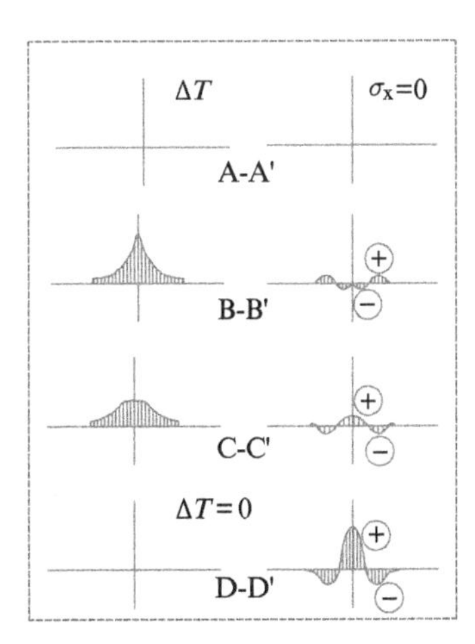

Figure 2.12 Residual stress variation along heat moving path.

- The weld area is heated up strongly in the welding procedure, which produces a high temperature difference between the fusion area and the region surrounding the weld.
- The heated area expands due to temperature difference with surrounding area, causing thermal stress. The yield stress can be exceeded in some parts and the weld fusion zone is plastically compressed.
- The temperature difference is gradually removed while the residual stress is formed in the cooling-down process. During the cooling process, the weld pulls on the surrounding solidified material, which resists, inducing tensile residual stress in the weld area and compressive residual stress in the surrounding region.

There are three types classified residual stress according to magnitude, as shown in Figure 2.13 (Anderson 2000):

- Type I (σ_I): first order residual stress, which refers to macro residual stresses that develop in the body of a component or structure on a scale that extends over macroscopic areas and that are averaged over several grains. It is of particular relevance for engineering purposes and therefore is the main focus of this research.
- Type II (σ_{II}): second order residual stresses, which are defined as micro residual stresses varying on the scale of an individual grain. Such stresses may be expected to exist in single-phase materials and this type of residual stresses act in approximately 0.01–1 mm.
- Type III (σ_{III}): third order residual stresses that are micro residual stress within a grain and act between atomic areas. They exist essentially as a result of the presence of dislocation and other crystalline defects in size of approximately of 10^{-6}–10^{-2} mm. Second order and third order residual stresses are frequently grouped together as micro stresses.

2.2.4.2 Generation of Welding Residual Stress

The rate of heat conduction through a medium in a specified direction is proportional to the temperature difference across the medium and the area vertical in relation to the direction of the heat transfer, but inversely proportional to the distance in that direction. It can be expressed by Fourier's law of heat conduction:

$$q_x = -kA\frac{\partial T}{\partial x} \tag{2.3}$$

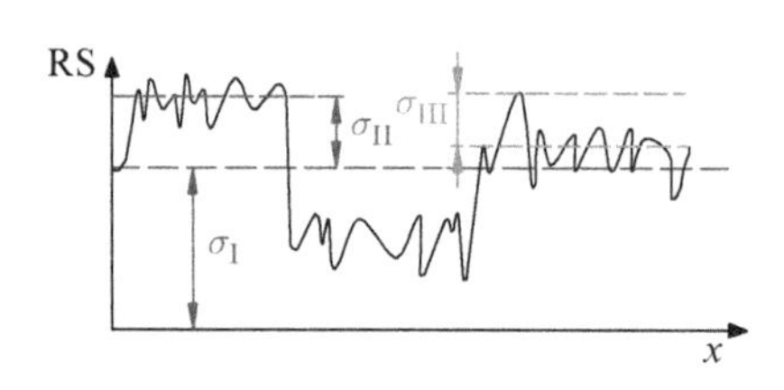

Figure 2.13 Three types of residual stress.

Where, q is heat flow (W), A is area perpendicular to heat flow (m^2), $\partial T / \partial x$ is temperature gradient in the direction of heat flow (°C/m), and k is thermal conductivity (W/ (m °C)).

Considering for an arbitrary control volume (Figure 2.14), the sum of the net energy of heat entering across the surface and the rate of energy generated in

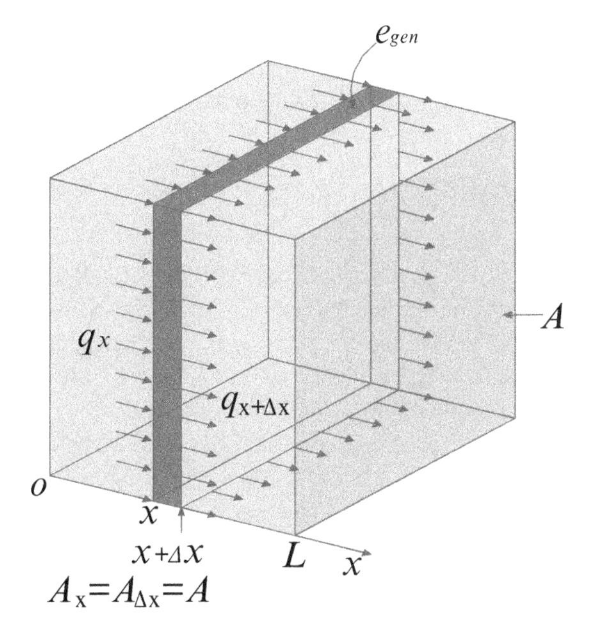

Figure 2.14 Heat transfer analysis in the welding process.

the control volume should be equal to the range of increase of internal energy of the control volume. According to this energy balance principle, for the one-dimensional heat conduction through a volume element, the energy balance on the element during a small time interval Δt can be expressed as:

$$q_x - q_{x+\Delta x} + \dot{e}_{gen} A\Delta x = \rho c \Delta x \frac{T_{x+\Delta x} - T_x}{\Delta t} \tag{2.4}$$

Therefore, by combining the Eq. 2.3 and Eq. 2.4, the one-dimensional transient heat conduction equation in a plane wall can be expressed as:

$$\frac{\partial}{\partial x}(k\frac{\partial T}{\partial x}) + \dot{e}_{gen} = \rho c \frac{\partial T}{\partial t} \tag{2.5}$$

In order to satisfy the residual strain compatibility relations, Eq. 2.6 can be obtained by differentiation. By doing this, a criterion judging whether the residual stress would be introduced can be defined, which is expressed as Eq. 2.7. If $R = 0$, it means that the plastic strain is the linear function of the coordinates and the deformation gradient is a constant. In this case, the deformation in the body is compatible. If $R \neq 0$, it means that the strain in the body is incompatible and the residual stress is produced due to the incompatible strain.

$$\left(\frac{\partial^2 \varepsilon_{x1}}{\partial y^2} + \frac{\partial^2 \varepsilon_{y1}}{\partial x^2} - \frac{\partial^2 \gamma_{xy1}}{\partial x \partial y}\right) + \left(\frac{\partial^2 \varepsilon_{x2}}{\partial y^2} + \frac{\partial^2 \varepsilon_{y2}}{\partial x^2} - \frac{\partial^2 \gamma_{xy2}}{\partial x \partial y}\right) = 0 \tag{2.6}$$

$$R = -\left(\frac{\partial^2 \varepsilon_{x2}}{\partial y^2} + \frac{\partial^2 \varepsilon_{y2}}{\partial x^2} - \frac{\partial^2 \gamma_{xy2}}{\partial x \partial y} \right) \tag{2.7}$$

$$\frac{1}{E}\left(\frac{\partial^4 F}{\partial x^4} + 2\frac{\partial^4 F}{\partial x^2 y^2} + \frac{\partial^4 F}{\partial y^4} \right) = R(x, y) \tag{2.8}$$

Importing Eq. 2.7 into Eq. 2.6, Eq. 2.8 can be obtained to be the criterion for judging whether the residual stress is introduced or not. Eq. 2.8 is a four-order non-homogeneous differential equation and the residual stress can be obtained when the solution of the equation can be found. In practice, it is hard to find the solution of Eq. 2.8 because of the complexity of the boundary conditions, which cause the numerical method to become a common tool for the thermal-stress analysis. In order to understand the formation mechanism of residual stress during the welding, a three-bar model is proposed to analyze the influence of thermal cycles on residual stress.

2.2.5 Residual Stress Investigation Techniques

2.2.5.1 Experimental Investigation

Some residual stress measurement techniques can be found at this moment. Table 2.3 gives the details about the comparison for these techniques (Kandil et al. 2001). Among the listed techniques, hole-drilling is a widely used method for measuring residual stress and was first proposed by Mather (1934). Rendler and Vigness (1966) developed it into a systematic procedure. To simplify the calculation procedure for non-uniform residual stress fields, Schajer (1988a, 2009) developed the appropriate calibration coefficients for the incremental hole-drilling method. Pang (1989), Payne (Payne and Porter-Goff 1986), Clarin (2004), Anderson (2000), and Acevedo (2009, 2011) all used this method to measure the welding-induced residual stress distribution for different kinds of weld joints. Some valuable reviews about the application of the hole-drilling method can be found (Grant et al. 2002; Micro-Measurements 2007; Oettel 2000). To standardize the operations and calculation procedures of the hole-drilling method, ASTM E837-08 (ASTM 2008) was formed to determine residual stress near the surface of an isotropic linear-elastic material.

Another technique frequently used is X-ray and neutron diffraction (Claphama et al. 2004; Fitzpatrick et al. 2005; ISO/TS 2008; Prevéy 1996). It relies on the elastic deformations within a polycrystalline material to measure internal stresses in a material (Fitzpatrick et al. 2005). It is a non-destructive technique, however, the limitation imposed on the test specimen size and geometry causes restrictions in application. It can be used to measure the residual stress localized on the surface. Neutron diffraction has the same work principle as x-ray diffraction (Withers 2007). However, the greatest advantage that neutron has over x-ray is the very large penetration depths. Neutron diffraction can also provide complete three-dimensional stress maps of a material. The

Table 2.3 Comparison for testing methods of residual stress.

Technique	Physical characteristics			Advantages and disadvantages
	Penetration	**Stress state**	**Destructive**	
Hole-drilling	Hole diameter	Uniaxial Biaxial	Semi	• Quick, simple • Widely available • Portable • Wide range of materials • Destructive • Limited in sensitivity and resolution
X ray diffraction	1 mm	Uniaxial Biaxial	No	• Widely available • Wide range of materials • Lab-based system • Limited in sample size
Synchrotron	>500μm	Uniaxial Biaxial Tri-axial	No	• Improved penetration of x-rays • Depth profiling • Lab-based system • Specialist facility only
Neutron diffraction	25 mm	Uniaxial Biaxial Tri-axial	No	• Excellent penetration and resolution • 3D maps • Lab-based system • Specialist facility only
Curvature and layer removal	Not applicable	Uniaxial Biaxial	Yes	• Simple to handle • Wide range of material • Limited to simple shapes • Destructive
Magnetic	20–300μm	Uniaxial Biaxial	No	• Simple to handle • Wide range of material • Portable • Only suitable to ferromagnetic materials
Ultrasonic	>100 mm	Uniaxial Biaxial	No	• Simple to handle • Portable • Low cost • Fast • Limited resolution

successful application of neutron diffraction can be found in some researchers' work, such as Claphama et al. (2004), Withers (2007), and Webster (2001). Brule and Kirstein (2006) give the introduction of the residual stress diffractometer KOWARI and its application in many projects.

2.2.5.2 Numerical Modeling

Finite element simulation is another powerful tool to investigate the formation process of residual stress. Large progress on the simulation of welding residual stress has been made during last decades. Welding is a complicated process that involves some couplings, and therefore, it is necessary to simplify this process in the actual modeling. Figure 2.15 shows the relationship between the heat flow, deformation, and metal microstructure in the welding and Table 2.4 lists the phenomena in the welding process (Lindgren 2001b). However, most modeling investigations on residual stress are based on the 2D sequentially-coupled analysis that is organized in two steps. Some of couplings are ignored in this analysis. In this way, the thermal analysis is performed and the temperature field is obtained. After that, the mechanical analysis is performed by importing the temperature field as the thermal loading.

Currently, 2D models are still dominating the published works. This can be ascribed to the fact that 3D modeling needs a large amount of computing effort. Furthermore, most of those models are based on plane strain condition. This means a thin slice perpendicular to the motion of the source is selected and the longitudinal deformation is ignored in the analyses. In the process of welding simulation, a common concern in the welding residual stress modeling is to account for the interaction effect between the welding parameters, the evolution of material microstructure, temperature field, and residual stress field (Lindgren 2001a). It is therefore necessary to consider the impact of material properties' change during the welding on the final residual stress field.

The work explored by Lindgren and Karlsson (1988), who used shell elements to model a thin-walled pipe, is a pioneer work in the 3D residual stress simulation. Compared with 2D models, 3D models can give more accurate results of residual stress fields because they can include all strains and stress components. Ueda et al. (Ueda and Nakacha 1982; Ueda et al. 1976; Ueda et al. 1993) and Wang et al. (1996) gave explorations on the three-dimensional residual stress analysis for a pipe-plate joint. Wu et al. (1996) described the 3D modeling procedure of the welding for a T-joint. They created the 3D model as the reference model and established 2D models to perform parametric studies. Lindgren (2001a; 2001b) gave the modelling details for residual stress field.

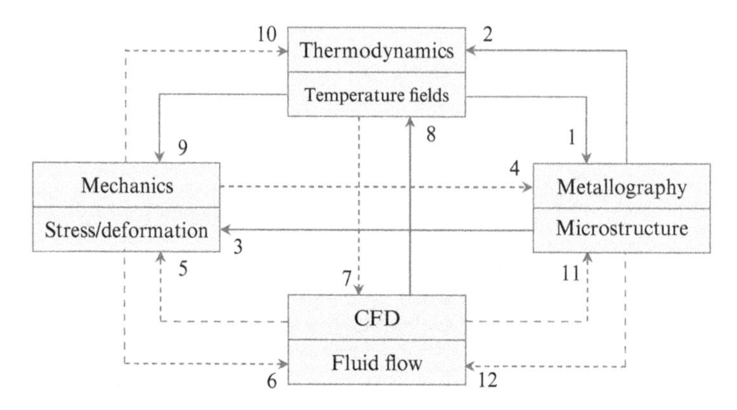

Figure 2.15 Couplings in residual stress formation.

Table 2.4 Couplings happened in the welding process.

Coupling	Description of coupling	Nature
1	Microstructure significantly depends upon temperature	Strong
2a	Latent heat is released during the phase transformation	Medium
2b	Microstructure affects the thermal conductivity and heat capacity	Medium
3a	Phase change results in associated plastic strain	Strong
3b	Thermal expansion depends on microstructure	Strong
3c	Elastic/plastic material behavior depends on microstructure	Strong
4	Phase changes are affected by the stress state	Weak
5	Flow patterns are affected by structural deformations	Weak
6	Fluid pressure produces deformation	Weak
7	Fluid velocity is affected by temperature	Weak
8	Fluid flow produces enhanced heat transfer	Strong
9	Fluid flow produces enhanced heat transfer	Strong
10	Heat is generated due to mechanical deformation (due to elastic, plastic, and thermal strain rate)	Weak
11	Material behavior is altered by the flow pattern	Weak
12	Material behavior affect flow pattern	Weak

One complication of the residual stress modeling is the simulation for multi-pass welding, which is the most frequently used in the practice. In order to reduce the computational effort, lumping technique is frequently used. Ueda et al. (Ueda and Nakacha 1982) first studied the multi-pass welding. The weld passes were lumped into different numbers of blocks. It was concluded that the lumping method is an effective way to get an accurate result. The influence of ignoring some weld passes on the residual stress field for a narrow gap welding is investigated. It is concluded that only the final weld pass should be modeled if the residual stresses on the surface were of primary interest. Rybicki (Rybicki et al. 1978; Rybicki and Stonesifer 1979), Free, Brickstad (Brickstad and Josefson 1998a), and Hong (Hong et al. 1998) all tried different lumping skills in the modeling. Another way to reduce the computation cost is by translating and superimposing the residual stress field of a single pass into multi-pass.

Another complication of the residual stress modeling concerns the addition of filler material. Two methods can be found to handle this. One way is to include the whole structure in the computation model from the beginning, including the weld that has not been laid yet. However, the elements corresponding to the weld that has not been laid yet are set with reduced material properties to eliminate the influence of those elements on the rest of model. Those elements corresponding to the weld that has not been laid yet are called "quite elements" (Brickstad and Josefson 1998b). Rybicki (Rybicki et al. 1978; Rybicki and Stonesifer 1979), Brust (Brust and Rybicki 1981), and Michaleris (1996) used this method in their models. Another way is to restructure the model at different times. The elements corresponding to the weld that has not been laid yet are not included in the analysis and they are nominated as "inactive elements". This is commented as being a more accurate method, but it requires applicable facility of finite element packages and user interaction at each new weld pass.

Some investigations on residual stress field of plate and tubular structure can be found. Jang et al. (2007) analyzed the residual stress distributions in welded T-joints with a three-dimensional non-steady heat conduction analysis and thermal elastic-plastic analysis. Lee et al. (2006) investigated the magnitude of residual stress and the effect of residual stress on component failure for T-plates and tubular T-joints. Residual stress distribution was measured along the thickness direction from weld toe and the neutron diffraction method was used to measure the actual stresses along the thickness direction. Two kinds of plates were introduced to fabrication T-plate samples. One was medium-strength steel with a yield strength of 348 MPa and the other had a yield strength of 700 MPa. The tubular joints were made of grade 355 steel. It was shown that the ratio of residual stress to yielding stress in transverse direction of weld toe is 0.15 around, but this value increases to 0.6 rapidly at the position where the ratio of distance from plate surface to plate thickness is 0.1. For tubular joints, it was shown that the maximum residual stress will turn out in crown and the magnitude can be 0.48 of yielding stress. Recently, Acevedo and Nussbaumer (2010) studied the residual stress distribution of welded tubular K-joints near the weld toe and its impact on the crack growth. They assumed that when the ASTM hole-drilling strain gauge method is employed for residual stress measurement, the residual stress distribution measured near the weld toe will not be affected even when the brace of the joint was cut off to facilitate the measurement procedure.

2.2.6 Exploration on Residual Stress Effects

In the welding process, thermally generated residual stresses on a macroscopic level are often the consequence of non-uniform heating or cooling which may lead to severe thermal gradients that produce large internal stress. The residual stresses due to welding in HSS could do harm to the integrity of structures. Residual stress not only affects the initiation and onset of the propagation of surface cracks but also changes the path of a crack as it grows below the surface. It is commented (Clarin 2004) that the effect of residual welding stresses on the performance of a welded structure is particularly significant when low stresses are applied. It is proposed (Maddox 1991) that the magnitude of residual stresses increase with increased material yield strength and a higher residual stress produced in welded high-strength steel could lead to a reduction in fatigue strength when compared with those for lower ones. Stacey and Barthdemy (Stacey et al. 2000; Stacey et al. 1996) incorporated the effects of residual stress into their structural integrity assessment procedure. Billingham et al. (Billingham et al. 2003; Billingham and Spurrier 1995) and Bjorhovde (2004) summarized some aspects for the use of HSS and gave a brief discussion on the influence of residual stress.

It was found by Krebs et al. (2007) that, in welded components, there is a considerable decrease of residual stresses as an effect of external loading (Figure 2.16). When residual stress exists, the same slopes of S–N curves can be used to predict the fatigue life of the components. However, it is found that fatigue strength of welded components is significantly dependent on mean stress, whereas it is recommended that *the fatigue strength is regarded dependent on stress range only and independent of mean stress and stress ratio*. A new stability of residual stress would be attained after applying an external loading, and this affects as mean stress.

To find the impact of loading case on the fatigue performance of HSS, Sonsino (Sonsino et al. 2004) checked damage accumulation of welded butt and transverse stiffeners made of medium- and high-strength steels under fully reversed and pulsating loading. It is indicated that mean stresses do not affect the fatigue strength of medium- and high-strength steel if high-tensile residual stresses are already present. The removal of tensile residual stresses by thermal stress relief for welding in high-strength steel is effective in the range of high-cycle fatigue while this stress relief can be dropped within low-cycle fatigue range as cyclic plastic deformation at the crack-like weld toes will have eliminated the tensile residual stresses already (Sonsino 2009).

Similar work is carried out by Kaufmann (Kaufmann et al. 2007), who has also introduced the factor of real damage sums in research in order to give recommendations for the fatigue life estimation and concluded that no significant difference in the bearable local stress amplitudes for butt welds can be detected for the four investigated steels (S355N, S355M, S690Q, and S960Q) under constant amplitude loading. However, the butt welded of high-strength steel S960Q has advantages in the case of the normal Gaussian spectrum and in the case of overloads, especially under pulsating loading. Only slight advantages for the HSS S960Q exist in the transverse welding under pulsating overloads. Fatigue life prediction of offshore high-strength steel structures under stochastic loading was investigated (Pedersen and Agerskov 1991).

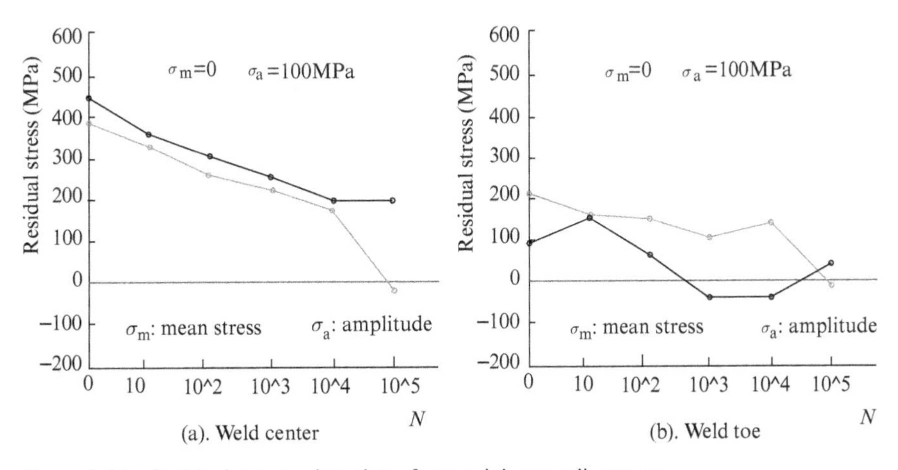

Figure 2.16 Residual stress relaxation after applying tensile stress.

Ravi et al. (2004) has analyzed and summarized those influencing factors on fatigue behavior of HSS butt joints, such as mis-match ratio, stress intensify factor range, notch location, and post-weld heat treatment. As to mis-match ratio, fatigue failure occurs much faster at higher stress intensity factor range when mis-match ratio is less than 0.8. It becomes very difficult to fabricate the joints without any weld defects when mis-match ratio is greater than 1.2 because of the large difference in thermal expansion coefficients existing between weld metal and base metal. It was concluded that the maximum fatigue life value can be attained in the combined conditions with high level of mis-match ratio, notch in weld metal rather than heat affecting zone and post-weld heat treatment. Total fatigue life of the over-matched joints is greater than the under-matched joints because of superior mechanical properties of the weld metals, ideal micro-structure of the weld regions, and favorable residual stress patterns in the weld metal region.

Agerskov (Agerskov 2000; Agerskov and Petersen 1998) carried out the comparative investigation on fatigue performance of welded joints made of HSS and mild steel in constant amplitude and variable amplitude by full-scale a offshore tubular test and a fracture mechanics analysis. It was found that the validity of Miner's law will depend on the load history in tension and compression by investigating hollow section connections such as fillet weld and longitudinal and transverse butt weld and hollow section joints including X-joints and K-joints made of Fe E 460 and Fe E 600.

Etube (2001) has presented details of variable amplitude fatigue tests conducted on full-scale circler tubular welded Y-joints fabricated from 700 MPa yield strength steel-SE702 which is a typical HSS used in construction of jack-up rig legs by applying out-of-plane bending loads. By comparing the tests results with data on mild steel from previous studies, it was found that tubular joints made of HSS are at least as good as conventional steels and may offer the benefit of longer fatigue lives at lower stress levels. It also accepted the fact that HSSs are more susceptible to corrosion fatigue and hydrogen-induced stress corrosion cracking. It is shown by King (1998) that the susceptibility to stress corrosion cracking increases with increasing yield stress and HSS might show inferior properties in seawater with highly negative cathodic protection potential.

It is found that the residual stress due to welding has a great influence on the crack growth rate in steel structure. The coefficient C, which is used in the Paris law ($\frac{dA}{dN} = C\Delta K^m$), varies as a function of the magnitude and distribution of the residual stress in the welded structure (Figure 2.17). A new coefficient is proposed to include the residual stress effect when evaluating the fatigue performance of the structure:

$$C^R = C\frac{1+\dfrac{K_{\mathrm{res}}}{K_{\max}}}{1+\beta\dfrac{K_{\mathrm{res}}}{K_{\max}}} \tag{2.9}$$

where K_{min} is the minimum stress intensity factor, K_{max} is the maximum stress intensity factor, and K_{res} is the stress intensity factor due to the welding residual stress which can be determined by Green's function.

2.3 Fatigue Analysis of Tubular Joints

2.3.1 Classification and Parameters of Tubular Joints

Box hollow section (BHS) joints have sparked extensive interest in supporting structures in offshore engineering and onshore structures. A typical BHS joint is made up of one or several smaller rectangular tubes called braces, and one larger rectangular tube called chord. Unlike the traditional rectangular hollow section and circular hollow section joints, which are made of mild steel, the braces and chord in a BHS made of HSS are formed by welding rather than casting. This is because HSS cannot suffer such great bending to form components and is thus only available in the form of a sheet. After the braces and chord are formed, the braces are jointed with the chord by welding the braces onto the outer surface of the chord.

The connection between the braces and the chord is the most critical part of the entire structural system. The box intersection is simpler when compared with CHS joints since it only needs straight cuts, which is one of the reasons why box tube and rectangular tube joints are widely used in engineering. The BHS joints can be classified into many types according to their geometries. Using this classification, BHS joints can be classified as T-joints, Y-joints, K-joints, X-joints, TT-joints, XX-joints, TX-joints, and KK-joints. The configurations of these common hollow section joints are illustrated in the Figure 2.18. Although other classifications, such as by the loading path, have also been proposed, geometry classification is adopted frequently because it is easier to understand for designers.

For brevity in design and analysis, the geometry of a BHS joint can be simplified by expressing it as non-dimensional parameters. Figure 2.19 shows the definition of these parameters that are commonly used in practice. Hence, the characteristics of a K-joint can be described by several normalized geometrical parameters, such as α, β, γ, τ, θ, and ζ.

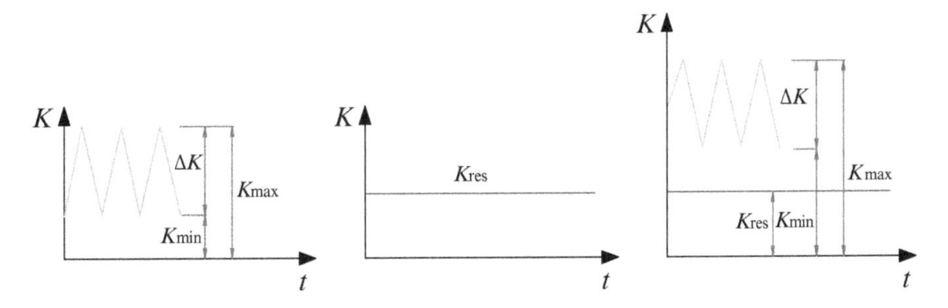

Figure 2.17 Impact of residual stress on K value.

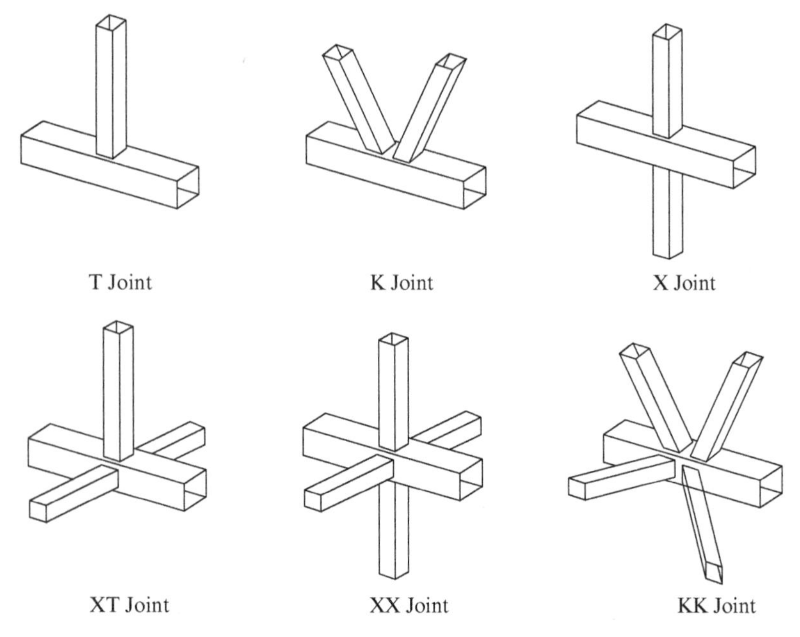

Figure 2.18 Common rectangular tubular joint configurations.

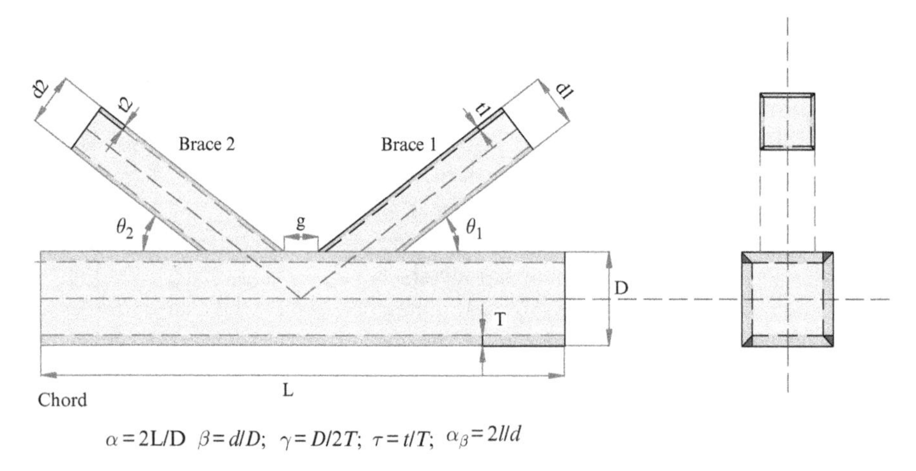

Figure 2.19 Geometrical parameters for tubular joints.

2.3.2 Stress Analysis of Intact Tubular Joints

Accurate understanding of the distribution of stress concentration around welded intersections is of great importance since it plays a key role in the prediction of fatigue life. The stress-life (S–N) curve approach is conventionally used in the process of fatigue design of the hollow section joints (Figure 2.20). The fatigue life of the hollow section joints can be predicted based on the hot-spot stress range at the welded intersection, which has been used together with parametric stress equations

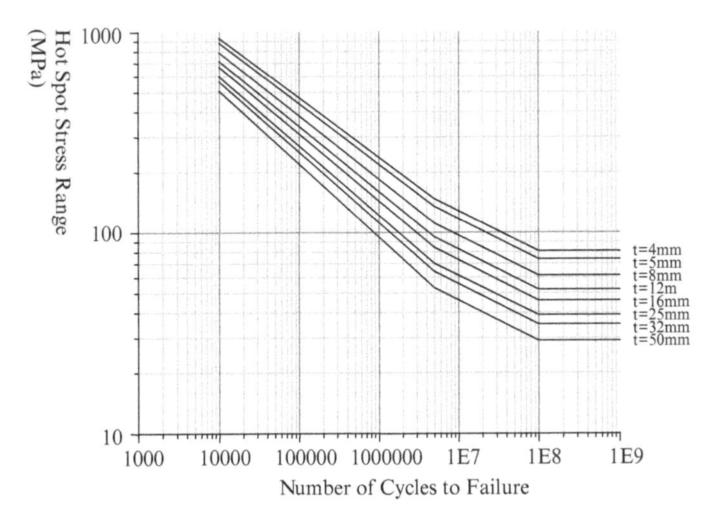

Figure 2.20 Typical S–N curves for fatigue design of CHS and RHS joints.

to ensure that the hollow section joints satisfy the fatigue design requirements for many years already. On the other hand, when the spot stresses are measured by experimental techniques and numerical modeling in past research, the residual stresses caused by welding are eliminated and missed in analysis. But when joints are made of HSS, these stresses have significant influence on the fatigue life of the joints, so they cannot be ignored.

Therefore, not only is information required on the absolute maximum value of stress at the intersection, but also the distribution of stress around the intersection and the stress variation through the thickness must be available under different types of load conditions. However, stress distributions in welded hollow section joints are very complicated due to the non-uniform stiffness behavior around the intersection of brace and chord, the structural discontinuities, and welding. Therefore, according to the cause of the stress, the stresses in the intersection of a hollow section joint may mainly include: nominal stress, hot-spot stress, local or notch stress, and residual stress. Therefore, some fatigue analysis procedures are produced according to different stress bases.

- Nominal Stress

This is the basic stress of the hollow section member due to the applied load. It can be calculated by simple elastic beam theory. The following equation can be used to calculate this stress.

$$\sigma_n = \frac{F}{A} + \frac{M}{I} \cdot y \tag{2.10}$$

Where, F is axial load, A is area of cross section, M is bending moment, I is the moment of inertia of the cross section, and y is the distance from the neutral axis to the point considered.

- Hot-Spot Stress The concept of hot-spot strain as a useful design parameter for low-cycle fatigue and fracture was initially developed during the mid of 1960s. The hot-spot stress and stress concentration factor are obtained by strain-based measurement and calculation since this method has its intrinsic advantages. One is that strain can be measured easily by individual strain gauge, while strain rosettes are needed when the definitions are stress-based. The other reason is that stresses cannot significantly exceed the yield stress. An almost horizontal S–N line, which does not occur, would be produced if a relatively constant number of cycles to failure would result for strain ranges larger than twice the yield strain.

One significant issue about the hot-spot stress method is the use of principal stress or a stress perpendicular to the weld toe. In other words, the definition of hot-spot stress is still obscure and debatable. The definitions in DEn (1990) and EC3 (1993) apply principal stress while IIW (1999), AWS (2008), and API (1993) concern a stress perpendicular to the weld toe.

DEn in its design guidance recommended that hot-spot stress can be given as the peak value of geometric stress found at the weld toe and it should incorporate the effects of overall tube geometry (i.e., the relative size of the brace and chord) but omit the concentrating influence of the weld itself, which results in a local stress distribution.

In view of van Wingerde (van Wingerde et al. 1995), the application of stress perpendicular to the weld toe is preferable. This is because the principal stresses will be much higher than stresses perpendicular to the weld toe, but this difference decreases with proximity to the weld toe. This is shown in Figure 2.21, in which stress perpendicular to the weld toe is considered for the hot-spot stress in the chord and braces of tubular joint. Another significant reason is that only stress components perpendicular to the weld toe significantly relate to weld shape. Furthermore, simple strain gauges can be used as a substitute of gauge rosette when only strain

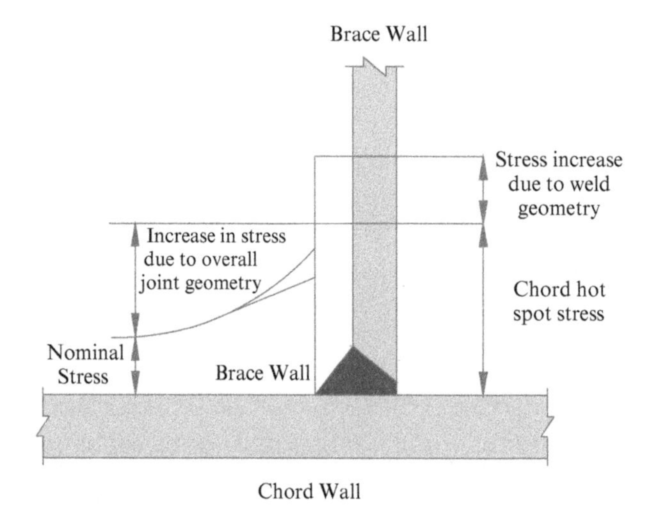

Figure 2.21 Definition of hot-spot stress.

perpendicular to the weld toe is to be measured. Finally, it would require extrapolating all components of the principal stress if this stress was applied, and the direction of the principal stress would be different for diverse loading cases, which makes it difficult to superimpose loadings.

Another debatable issue is the extrapolation procedure, which is applied to obtain the hot-spot stress to exclude the influence of the local notch stress in certain areas away from the weld toe. Linear extrapolation is frequently used to measure the stress perpendicular to the weld toe. It is the Working Group III of the ECSC who originally proposed a value of $0.2\sqrt{r \cdot t}$ (r is the radius and t is the wall thickness of the brace) as the minimum distance from welding. Later, Gurney and van Delft put forward that the minimum distance should be at least 4 mm, considering the influence of r on the extrapolation position.

In many design guides, hot-spot stresses are usually calculated in terms of stress concentration factors (SCFs). The SCF is the ratio between the hot-spot stress, caused by structural discontinuities or welding, and the nominal member stress. It is related to the geometry of the joints and the procedure of fabrication. Thus, the hot-spot stress calculations can be simplified to determine the SCFs for the certain hollow section joints. In practice, the hollow section joints are usually subjected to combined loadings. So, the hot-spot stress is a function of all nominal stress multiplied by their SCFs. In the case of Y and K connections where only axial forces and in-plane bending moments are considered, the total hot-spot stress can be determined by:

$$S_{rh,s} = \sigma_{rm1} SCF_{m1} + \sigma_{ra1} SCF_{a1} + \sigma_{rm0} SCF_{m0} + \sigma_{ra0} SCF_{a0} \tag{2.11}$$

where, σ_{rm1} is the nominal in-plane bending stress range in the brace, σ_{ra1} is the nominal axial stress range in the brace, σ_{rm0} is the nominal in-plane bending stress range in the chord, σ_{ra0} is the nominal axial stress range in the chord, and SCF_{m1}, SCF_{a1}, SCF_{m0}, and SCF_{a0} are the corresponding SCFs.

- Notch Stress

This stress is the result of the geometric discontinuity of the hollow section walls due to the weld. The influencing factors may include the local notch of the weld toe, the undercut, and lack of fusion. This stress may vary widely along the weld and is strongly dependent on the skill of the welder and the welding method. It is not feasible to give explicit correction factors for these influences, so the current practices of demanding a minimum quality of the weld toe would have to be maintained.

3

Experimental Investigation of Residual Stress for High-Strength Steel Plate-to-Plate Joints

3.1 Introduction

High-strength steel shows reduced capacity for strain hardening after the yielding compared with mild steel since the steel strengthening mechanism is used to increase the yield strength. Therefore, use of HSS in construction should be under caution especially in the case of cyclic loads. During steel structure construction, welding is frequently used for element connections such as beam-column joints. However, due to the highly localized, non-uniform, and transient heating in welding and the non-linearity of material properties under elevated temperature, residual stress will be introduced as a result of inhomogeneous deformation. For HSS structures, the impact of residual stress on fatigue performance, fracture, and structural stability should be taken into account during structural analysis and design.

Hole-drilling technique (Schajer 1988a; Schajer 1988b; Schajer 2009; ASTM 2008; Micro-Measurements 2007), X-ray (Fitzpatrick et al. 2005; Prevéy 1996), and neutron diffraction (Park et al. 2004) are efficient methods of experimental investigation. Though X-ray diffraction technique offers a non-destructive alternative, it has severe limitations such as bulk and complexity of the equipment and measurement for only very shallow surface layers. The hole-drilling method is a common method for measuring residual stress by removing localized stress and measuring strain relief in the adjacent material. The hole-drilling method is to measure residual stress near the metal surface by stress relaxation when a hole is drilled into the center of a rosette strain gauge. It has advantages of reliability, simplicity, convenient practical implementation, and therefore it is broadly used in civil and mechanical engineering.

In this chapter, an investigation on the residual stress distributions near the weld toe of HSS plate-to-plate T- and Y-joints is carried out. Two groups of specimens, corresponding to welding performed at ambient temperature and at a preheating temperature of 100 °C, are fabricated. The effects of preheating and joint geometry on the residual stress distribution near the weld toe are investigated by applying the standard ASTM hole-drilling method for residual stress measurement. Furthermore, a preliminary study is also performed to evaluate the influence of brace plate cutting on the residual stress distribution near the weld toe of the joints.

Welded High Strength Steel Structures: Welding Effects and Fatigue Performance, First Edition. Jin Jiang.

3.2 The Hole-Drilling Method and Specimen Details

3.2.1 The ASTM Hole-Drilling Method

The modern application of the hole-drilling method can be traced to 1966 when Rendler and Vigness (1966) developed the hole-drilling method into a systematic procedure and thereby formed the basis of ASTM E837 standard (ASTM 2008). Beaney and Procter (1974) improved the application of an air abrasive machine to eliminate the introduction of additional residual stress in the process of drilling. Flaman (1982) developed another practical stress-free drilling method with ultra-high speed. Schajer (1988a, 1988b) later gave a review on the history and progress of the hole-drilling method and indicated some promising directions for future developments. Lu (1996) gave procedure of measurement of residual stresses. On the theoretical side, Schajer (2009) carried out the primary finite element analysis of the hole-drilling method to list calibration coefficients in the formula of calculation. Shim et al. (1992) investigated the determination method of residual stresses in thick-section weldment.

The standard ASTM hole-drilling method (ASTM 2008) is a common way of measuring residual stress by removing localized stress and measuring strain relief in the adjacent material. It causes relatively little damage to the specimen and allows localized residual stress measurements. Measurement of residual stresses by the hole-drilling strain gage method is systematically given (Micro-Measurements 2007). The principle of the method is to release the localized stress by introducing a small hole into a residually stressed structure and thus changing the local strain on the testing surface. By comparing the strain at this point before and after hole-drilling, stress relaxation due to hole drilling can be determined. On the assumption that the material of the plate is homogenous and isotropic and its stress-strain curve is linear, the relieved strain at a point $P(R, \alpha)$ can be obtained by substituting the stress relaxation into Hooke's law (ASTM 2008):

$$\varepsilon = \frac{1+\nu}{E} \cdot (\bar{a}) \cdot \frac{\sigma_x + \sigma_y}{2} + \frac{1}{E} \cdot (\bar{b}) \cdot \frac{\sigma_x - \sigma_y}{2} \cos 2\alpha + \frac{1}{E} \cdot (\bar{b}) \cdot \tau_{xy} \cdot \sin 2\alpha \tag{3.1}$$

where, E is the Young's modulus, ν is the Poisson's ratio, ε is the relieved strain, σ_x and σ_y are the stresses in the x and the y directions respectively, α is the angle from the x-axis to the maximum principal stress (Figure 3.1), and $\bar{a}$ and $\bar{b}$ are calibration constants. In residual stress measurement, after the three groups of relieved strains are obtained by a strain rosette, σ_x, σ_y, and α can be determined.

The combined stresses for the measured strains ε_1, ε_2, and ε_3 can be expressed as the formula following (ASTM 2008):

$$\begin{cases} P = \dfrac{(\sigma_y + \sigma_x)}{2} = -\dfrac{Ep}{\bar{a}(1+\nu)} \\ Q = \dfrac{(\sigma_y - \sigma_x)}{2} = -\dfrac{Eq}{\bar{b}} \\ T = \tau_{xy} = -\dfrac{Et}{\bar{b}} \end{cases} \tag{3.2}$$

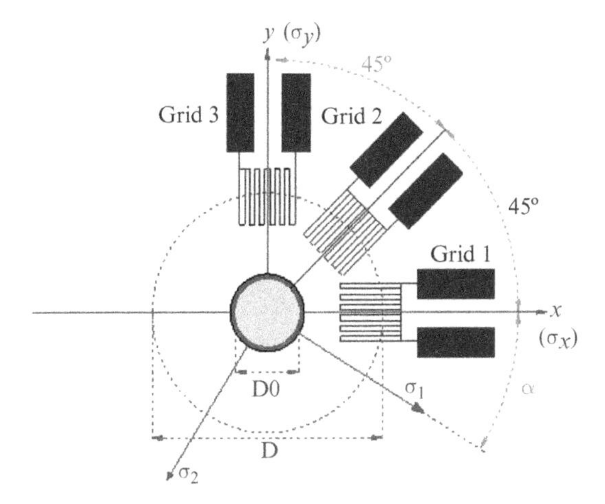

Figure 3.1 Schematic diagram of strain gauge for residual stress measurement.

where, p, q, and t are the three strain combinations with expressions of $p = (\varepsilon_3 + \varepsilon_1)/2$, $q = (\varepsilon_3 - \varepsilon_1)/2$, and $t = (\varepsilon_3 + \varepsilon_1 - 2\varepsilon_2)/2$, while E is Young's modulus and ν is Poisson's ratio. In this case, the in-plane Cartesian stresses can be expressed:

$$\begin{cases} \sigma_x = P - Q \\ \sigma_y = P + Q \\ \tau_{xy} = T \end{cases} \tag{3.3}$$

3.2.2 Specimen Specifications

In the present experimental investigation, a number of plate-to-plate T/Y-joints, made of HSS plate with minimum yielding stress of 690 MPa, were fabricated by SMAW. The RQT701 HSS plate used in this study was quenched and tempered structural steel with improved forming and welding performance by substituting some alloying elements with carbon. In order to prevent hydrogen cracking for the HSS plates, a moisture-resistant, ultra-low hydrogen welder that utilizes a covered electrode for low-temperature use, LB-70L, which is equivalent to the class AWS A5.5 E10016-G (AWS 2006) and supplied by Kobelco of Japan, was used. The diffusible hydrogen content of this electrode is 4ml/100mg. Table 3.1 gives the mechanical properties of the steel plate and the electrode. Table 3.2 lists the chemical composition of the RQT701 steel plate and the LB-70L electrode employed in this study.

As it is well known that preheating may significantly affect the residual stress distribution by delaying the cooling of the weld region and promoting hydrogen effusion. In order to compare the influence of preheating on residual stress distribution near the weld toe, two groups of specimens were prepared. In the first group, all welding steps were performed at ambient temperature (30 °C). All specimens from

Table 3.1 Mechanical properties of RQT701 steel plate and LB-70L electrode.

Items	Minimum yield strength (MPa)	Tensile strength (MPa)	Minimum average impact energy	Minimum elongation (%)
RQT701	690	790~930	27J@ -45 °C	18
LB-70L	685	755	108J@-60 °C	27

Table 3.2 Chemical composition of the RQT701 steel plate and the LB-70L electrode.

RQT701 Plate	C	Si	Mn	S	P	Cr	Mo	V
	0.14	0.40	1.35	0.003	0.012	0.01	0.12	0.05
	Ni	Cu	B	Al	Nb	Ti	O	-
	0.01	0.01	0.002	0.035	0.035	0.025	-	-
LB-70L electrode	C	Si	Mn	S	P	Cr	Mo	V
	0.04	0.40	1.28	0.003	0.006	0.04	0.48	0.01
	Ni	Cu	B	Al	Nb	Ti	O	-
	3.65	0.01	-	-	-	-	0.03	-

the second group were preheated before welding. Following the recommendation from the EN1011-2:2001 (BSI 2011), the minimum preheating temperature T_p is depended on the CE, the plate thickness d, the hydrogen content of the weld metal HD, and the heat input Q, and could be calculated as:

$$T_p = T_{pCET} + T_{pD} + T_{pHD} + T_{pQ}$$

$$\begin{cases} T_{pCET} = 750 \times CET - 150 \\ T_{pD} = 160 \times \tanh(d/35) - 110 \\ T_{pHD} = 62 \times HD^{0.35} - 100 \\ T_{pQ} = (53 \times CET - 32) \times Q - 53 \times CET + 32 \end{cases} \tag{3.4}$$

In Eq. 3.4, the unit of the temperature is in degree Celsius (°C). T_{pCET}, T_{pD}, T_{pHD}, and T_{pQ} are respectively the preheat temperatures corresponding to the equivalent carbon content, the plate thickness, the weld metal hydrogen content, and the heat input. By using the data given in Tables 3.1 and 3.2, the recommended minimum preheating temperatures for joints with 8 mm, 12 mm, and 16 mm RQT701 steel plates were found to be equal to 76 °C, 85 °C, and 94 °C, respectively. However, in order to make the fabrication procedure easy to handle, a single preheating temperature of 100 °C was applied to all preheated joints.

For each group of specimens, there are 6 different geometries, consisting of three different plate thicknesses (8 mm, 12 mm, and 16 mm) and two joint intersection angles (90° and 135°). All joints were fabricated by a qualified welder using hand welding. For the 90° joint, two duplicated specimens with the same welding procedure and joint geometry were fabricated in order to assess the human effects on residual stress distribution caused by SMAW. Hence, in the experimental study, a total of 18 joints were produced and tested. Figure 3.2 shows the typical specimen cross-sectional

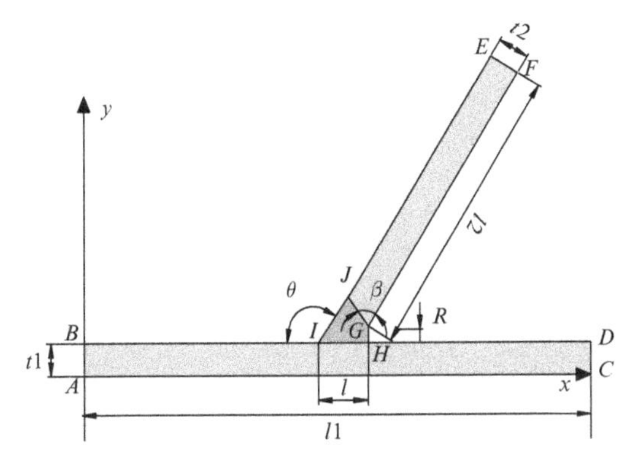

Figure 3.2 Typical welding profile of plate-to-plate joint (for θ = 90° and 135°).

geometry and the welding profile of the joint while Table 3.3 lists the geometry of the specimens. Note that the width of all specimens fabricated is equal to 150 mm. As shown in Figure 3.2, in a typical welding procedure, a T- or Y-joint was formed by adding the welding filler (region *I–J-G–H* in Figure 3.2) from the surface *H–G*, where it was packed by ceramic plate and ended at the surface *I–J*. The plate *A–B-C–D* and the plate *E–F-G–J* are named as the *chord plate* and the *brace plate*, respectively. θ and β are the joint intersection and the end preparation angles, respectively. l_1 and l_2 are respectively the length of chord plate and brace plate. t_1 and t_2 refer to the thickness of chord plate and brace plate, respectively. *R* is the weld root opening distance.

3.2.3 Welding Specifications

In this study, all full penetration welding profiles created for the plate-to-plate joints are complied with the AWS D1.1 2008 (AWS 2008) guideline. Table 3.4 lists the welding parameters for the fabrication. Before the welding started, the surfaces of the chord and brace plate were cleaned so that the specimens were free from slag and rust. This was then followed by end preparation. To avoid distortion and shrinkage, several splices were welded to assemble the chord and brace plates together. In the process of preheating, the area close to where the weld filler was added (Figures 3.3 and 3.4) was heated up to 100 °C. Temperature chalk was

Table 3.3 Geometry of the specimens (Note: width of all specimens = 150 mm).

θ (°)	$\beta\beta$ (°)	*L*1(mm)	*L*2(mm)	*W*(mm)	*t*1(mm)	*t*2(mm)	*R*(mm)	*L*(mm)
135	90	440	440	150	8	8	5	16
					12	12	5	22
					16	16	5	28
90	60	440	440	150	8	8	6	20
					12	12	6	27
					16	16	6	35

Table 3.4 Welding specification for the specimens.

		Heat treatment			Welding parameters		
Base metal	**Current**	**Preheat**	**Without preheat**	**Inter-pass temperature**	**Current**	**Voltage**	**Welding speed**
RQT701	DCEP	100 °C	30 °C	150 °C	170A	26V	2.6 mm/s

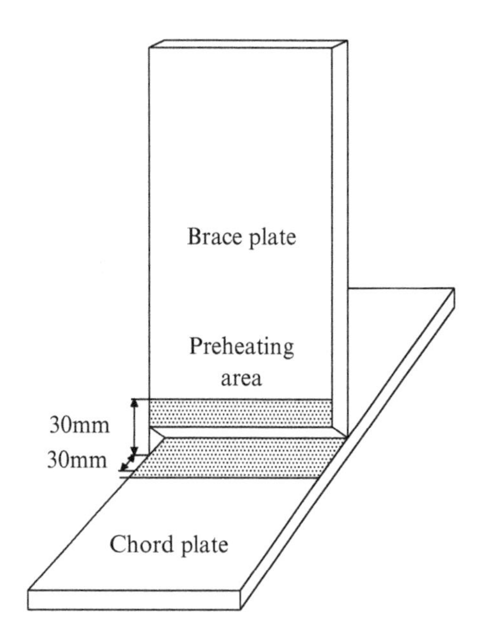

Figure 3.3 Preheating area.

Figure 3.4 Preheating process.

employed to ensure the preheating temperature was reached. For 135° joints, 9, 14, and 22 welding passes were employed for joints with plate thickness of 8 mm, 12 mm, and 16 mm, respectively. For 90° joints, 11, 13, and 17 welding passes were used for joints with plate thickness of 8 mm, 12 mm, and 16 mm, respectively. Figure 3.5 shows the number of weld passes and the welding sequences adopted for different joint angles and plate thickness. Given that the width of all specimens is equal to 150 mm, with an average welding speed of 2.6 mm/s, the time interval between each weld pass is roughly equal to 58s. It is found that the temperature in the weld filler after finishing a weld pass shall still be higher than 100 °C. Hence, for all specimens preheating was only conducted before the first weld pass and was not repeated between successive passes.

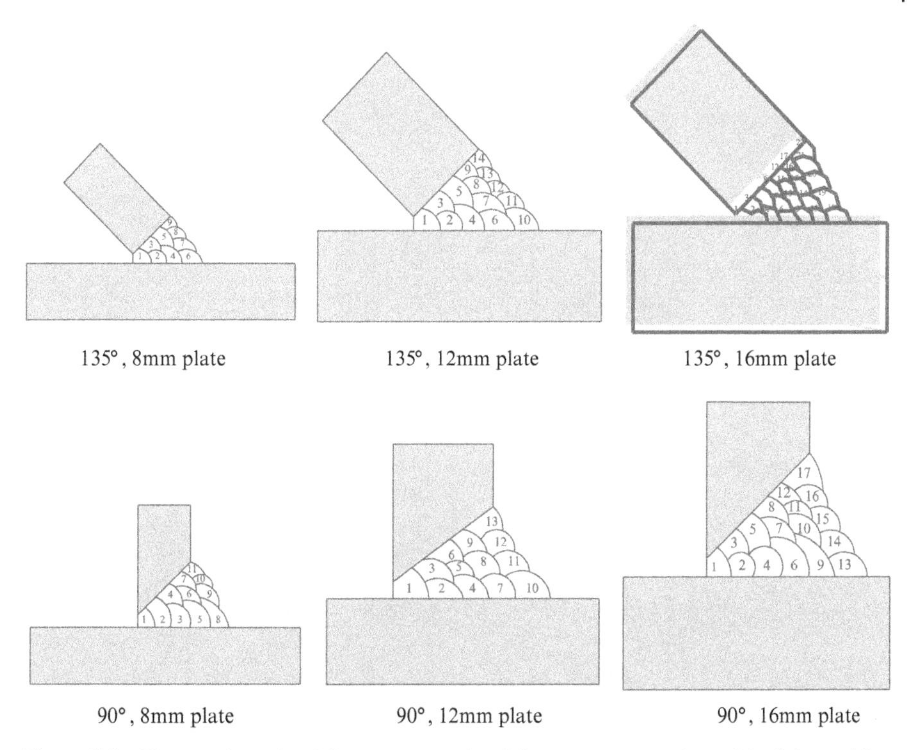

Figure 3.5 The number of welding passes and welding sequences adopted for joints with different thickness and intersection angles.

3.3 Residual Stress Investigation

3.3.1 Setup and Modification of the Hole-Drilling Guide

The RS-200 milling guide, a high-precision instrument for analyzing residual stress by the hole-drilling method through positioning and drilling of a hole in the center of a special strain gauge rosette, was used for measuring the residual stress in the specimens. The original form of the milling guide is shown in Figure 3.6(a). However, a physical limitation exists in the original setup: the journal hole fixing the high-speed air turbine is positioned at the center of the base of the milling guide. As a result, there exists a minimum offset distance of 50 mm between the edge of the guide base and the position of the drilled hole (center of the strain gauge). As the bracing plate of a 90° joint will obstruct the guide base, such constraint greatly handicaps the original guide's usage in this study. To overcome this problem, an improved set up was designed and fabricated as shown in Figure 3.6(b). In order to move the measurement point closer to the weld toe of the joint, the original triangular base was replaced by a tailor-made trapezoidal base supported by three swivel pads. Furthermore, the cylinder housing the journal was further machined down to reduce the minimum distance between the drilled hole and the guide base edge to 10 mm.

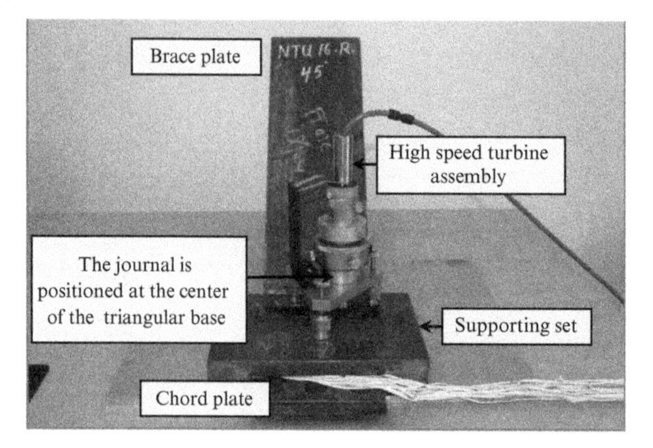

(a) The original milling

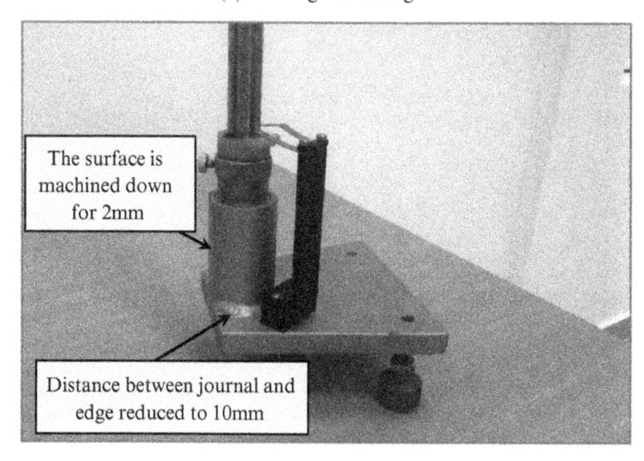

(b) The revised milling guide

Figure 3.6 The RS-milling guide for residual stress measurement.

3.3.2 Strain Gauge Locations

A special type of strain rosette FRAS-2, which is specially designed so that the three strain gauges (Grids 1, 2, and 3 in Figure 3.1) are all positioned on one side of the measurement point, was used to measure the released strain of the specimen during drilling. The strain gauge scheme employed for residual stress measurement on the chord plate is shown in Figure 3.7. For the 135° joints, along the longitudinal direction (the x axis in Figure 3.7), three strain gauges (A, B, and C) were placed at 5 mm from the weld toe (i.e., y =5 mm at Figure 3.7) with x =25 mm, 75 mm, and 125 mm, respectively. In addition, at the middle of the plate width where x =75 mm, another two strain gauges (B_1 and B_2) were placed at where y =20 mm and 35 mm, respectively. The strain gauge setting for the 90° was similar to the case of 135° joints, except that gauges A and C are removed and an additional gauge B_3 was placed at the middle of plate width with x = 50 mm.

The strain gauges scheme is slightly changed for 90° specimens because the residual stress measurement for plate-to-plate joints were carried out in two stages. The 135° specimens were tested first. Based on the test results of the 135° specimens, it was

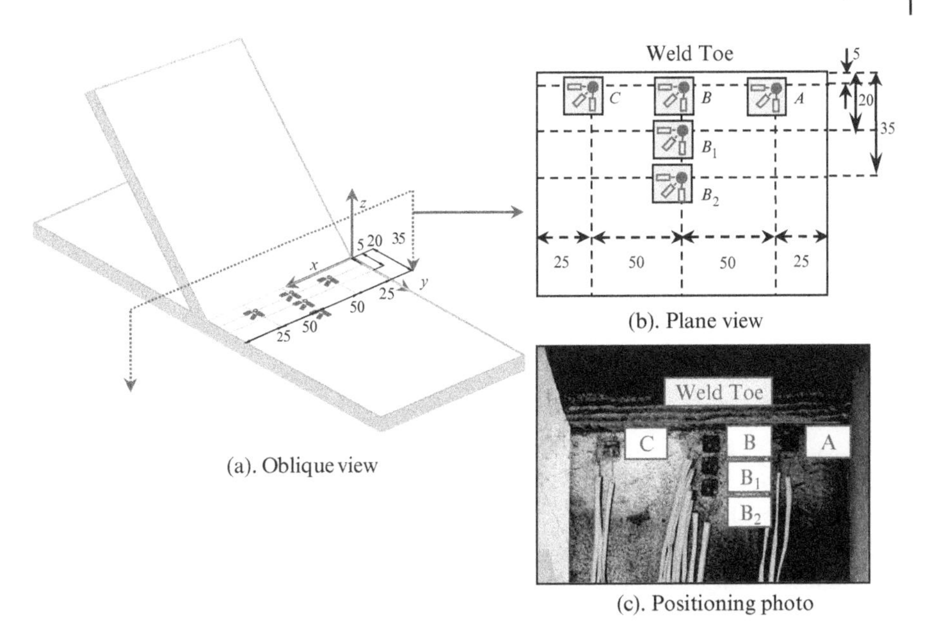

Figure 3.7 Strain gauge locations on chord plate for residual stress measurement (all dimensions in mm).

found the largest tensile residual stress is always engendered at the middle of the plate's width. For the residual stress at Points A and C, the magnitude of residual stress is much lower than that at Point B. In this case, for 90° specimens, the strain gauge at A and C is removed and another gauge is added along the width middle point so that the residual stress variation along the distance from the weld toe can be studied.

3.3.3 Calibration Test for Residual Stress Measurement

To ensure the accuracy of the residual stress measurement of the joints, a calibration test was performed to determine the coefficients $\bar{a}$ and $\bar{b}$ in Eq. 3.1. It was accomplished by using a small size (200 mm × 70 mm) calibration plate, which was cut out from the same batch of RQT701 steel plate that was used for joint fabrication. The calibration plate was fixed on a tensile machine and a strain gauge rosette was placed and oriented in such a way that Grid 1 and Grid 3 (Figure 3.1) were parallel to the loading direction and the transverse axis, respectively. Before hole-drilling, a small calibration loading P_c was applied to develop the desired uniform calibration stress σ_c which is much less than the yield stress of the plate. The values of relieved strain from Grid 1, ε_1', and Grid 3, ε_3', were then recorded. The loading was then released and the calibration plate was removed from the tensile machine. A standardized hole was drilled and the specimen was then remounted to the tensile machine. The same calibration loading P_c was re-applied and the corresponding Gird 1 and Grid 3 strains ε_1'' and ε_3'' were recorded. To make the results more reliable, eight groups of readings were recorded to obtain coefficients $\bar{a}$ and $\bar{b}$ that are required by the ASTM standard (ASTM 2008) at different depths ranging from 0 to

2 mm with 0.25 mm steps. Figure 3.8 shows calibration test for residual stress coefficients. Table 3.5 lists the values of the calibration coefficients obtained for different depths.

3.3.4 Residual Stress Measurement Procedure

For the residual stress measurement at the selected point, before mounting the strain gauge, a pair of crossed reference lines were first marked on the surface where strain measurements were needed. Careful alignment of the milling guide with the strain gauge hole center with a precision within 0.025 mm was accomplished by inserting a special-purpose microscope into the guide's center journal, and then positioning the guide precisely over the center of the rosette. A 1.6 mm diameter tungsten carbide cutter was installed in the turbine. The milling cutter was guided precisely during the drilling operation to ensure that it progressed in a straight line during drilling. Finally, a hole was bored by setting the compressed air supply to the high-speed turbine to produce a circular, straight-sided, and flat-bottomed hole. The hole first was advanced slowly to a depth of 0.05D (D is the diameter of gauge circle that is shown in Figure 3.1. The actual size of D is 5 mm) and the relieved strain was noted. The drilling and strain measurement steps were then repeated seven times with an incremental depth of 0.05D until a final hole depth approximately equal to

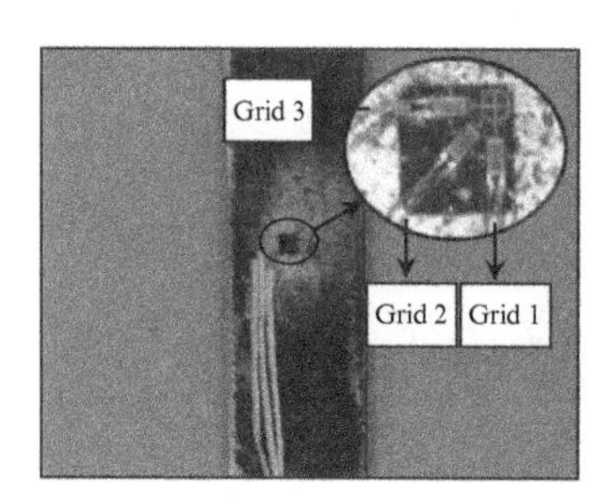

(a). Calibration testing specimen

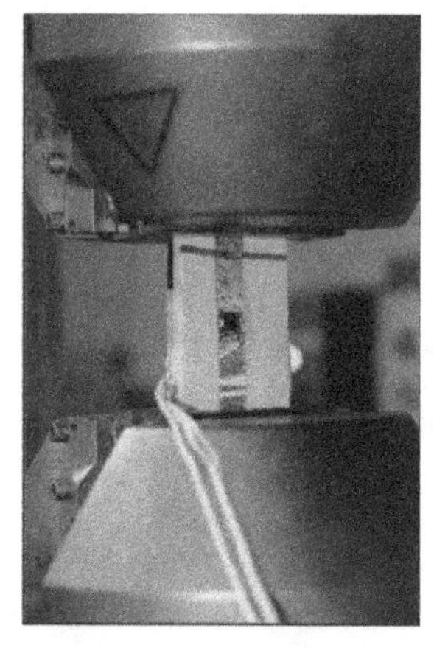

(b). Calibration test rig

Figure 3.8 Calibration test for residual stress coefficients.

Table 3.5 Coefficients $\bar{a}$ and $\bar{b}$ at different depths.

d (mm)	0.25	0.50	0.75	1.00	1.25	1.50	1.75	2.00
d/D	0.05	0.10	0.15	0.20	0.25	0.30	0.35	0.40
$\bar{a}$	0.031	0.069	0.095	0.121	0.118	0.142	0.142	0.149
$\bar{b}$	0.034	0.112	0.157	0.199	0.227	0.326	0.304	0.360

0.4D was achieved. The calculation procedure followed by ASTM E837-08 prefers averaging the measured 8 groups of data to reduce the effects of random strain measurement.

3.3.5 Cutting of Brace Plate

Despite the fact that in the current study the milling guide was modified to allow residual strain measurement to be carried out at locations as close as 5 mm from the weld toe of plate-to-plate joints with $\theta \geq 90$, the same set up is not applicable to joints with $\theta < 90$ or other joint types such as circular hollow section joints. In such cases, researchers may resolve to the method of brace cutting (Acevedo 2009) in which the brace plate obstructing the drill is removed to facilitate residual stress measurement near the weld toe. However, so far little information is available on the effect of the brace plate cutting operation on the residual stress distribution near the weld toe. Hence, in this study an investigation was conducted to study the effect of brace plate cutting. The brace plates of half of the 90° joints fabricated were mechanically cut before the residual stress measurements were conducted (Table 3.6, the fourth column). The cutting position was chosen to be a surface 5 mm above the weld toe at the brace plate. Figures 3.9 and 3.10 show a T/Y-joint before and after cutting, respectively.

3.4 Experimental Results

The measured residual stresses obtained from the experimental study are listed in Table 3.6. In the next five sections, results related to the distribution of residual stress along the weld toe, the effects of preheating, joint angle, plate thickness, and the brace-plate cutting effects are presented.

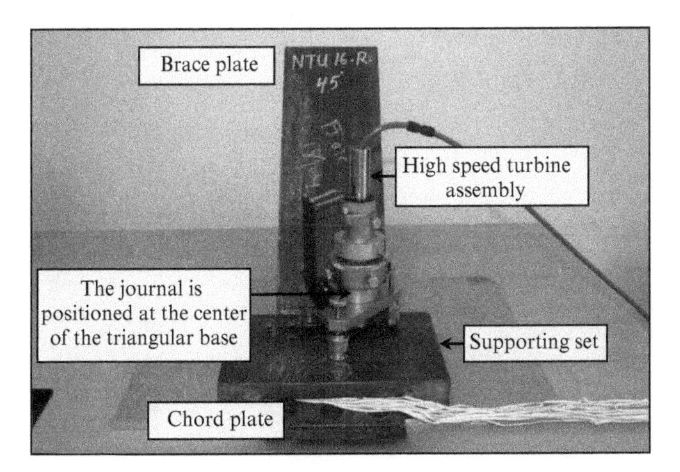

Figure 3.9 A T-joint before cutting of the brace plate.

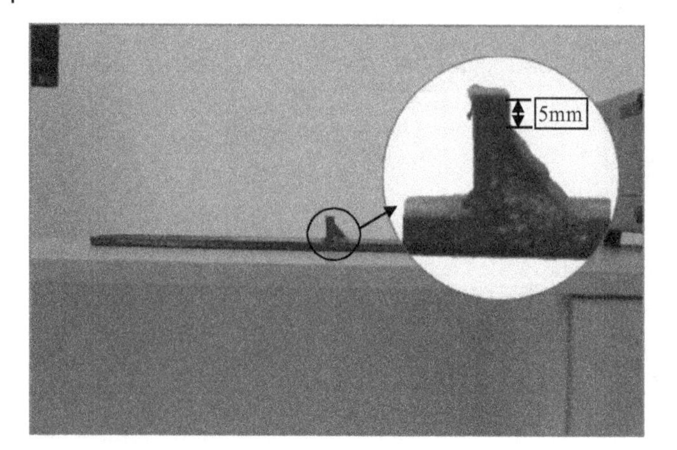

Figure 3.10 A T-joint after cutting of the brace plate.

Table 3.6 Residual stress measurement results.

Specimen cases				Stress in the measuring points (MPa) (-ve for compressive stress)					
θ (°)	t (mm)	Preheating	Brace plate cutting	A	B	B_1	B_2	B_3	C
90	8	Yes	Yes	–	−66.2	76.4	25.2	8.0	–
		No		–	−70.3	129.0	28.2	28.5	–
	12	Yes		–	−32.5	34.6	11.2	6.2	–
		No		–	−8.8	67.5	26.9	5.3	–
	16	Yes		–	51.1	91.9	25.2	5.2	–
		No		–	24.9	52.4	38.2	−26.4	–
	8	Yes	No	–	38.2	75.5	−16.0	−18.0	–
		No		–	110.1	60.8	65.7	−32.2	–
	12	Yes		–	59.3	42.4	−46.2	−20.4	–
		No		–	87.2	62.5	11.0	−42.6	–
	16	Yes		–	96.3	58.9	29.1	-6.3	–
		No		–	115.5	86.5	20.7	−18.3	–
135	8	Yes	No	−78.8	119.1	99.2	118	-	42.6
		No		23.2	151	96	90.9	-	−25.6
	12	Yes		−82.4	62.8	48.5	36.9	-	−68.2
		No		−15.6	102.6	79.7	32.3	-	−42.7
	16	Yes		47.7	132.8	114.9	53.4	-	16
		No		12	213.6	103.9	68.1	-	−68.5

3.4.1 Distribution of Residual Stress Along the Weld Toe

From the results shown in Table 3.6 for gauges *A*, *B*, and *C* for 135° joints, it can be seen that for both joints with and without preheat treatment, the residual stresses at the two ends of the joint (gauges A and C) are considerably less than that at the middle (gauge B). Furthermore, in some cases, even compressive residual stresses were recorded at gauges A and B. Such distribution of residual stress along the weld toe (*x* axis of Figure 3.7) is obviously due to the boundary effects at the two ends of the joint. Since the two ends of the joint were free from any in-plane restraint, the magnitude of the residual stress is much smaller than that at the middle of the joint. As a result, in the next four sections, focus will be given to the variations of residual stress along the transverse direction (*y* axis in Figure 3.7) at the middle of the joint (gauges B, B1, B2, and B3).

3.4.2 The Effects of Preheating

Figure 3.11 shows the residual stress distribution along the transverse direction (y axis) for 90° joints. Six curves are plotted to compare the preheating effect on residual stress distribution for different plate thicknesses. At gauge B (5 mm from weld toe), the maximum residual stress for joints without preheating treatment exceeded 100 MPa (roughly 1/6 of the yield strength). In addition, the residual stress for all preheated joints is less than that of their ambient counterparts (65.3%, 32%, and 16.6% reductions for 8 mm, 12 mm, and 16 mm joints, respectively). This stress-relieving benefit can still be observed at gauge B1 (20 mm from the weld toe) and gauge B2 (35 mm from weld toe). The only exception occurred for the 8 mm joint at

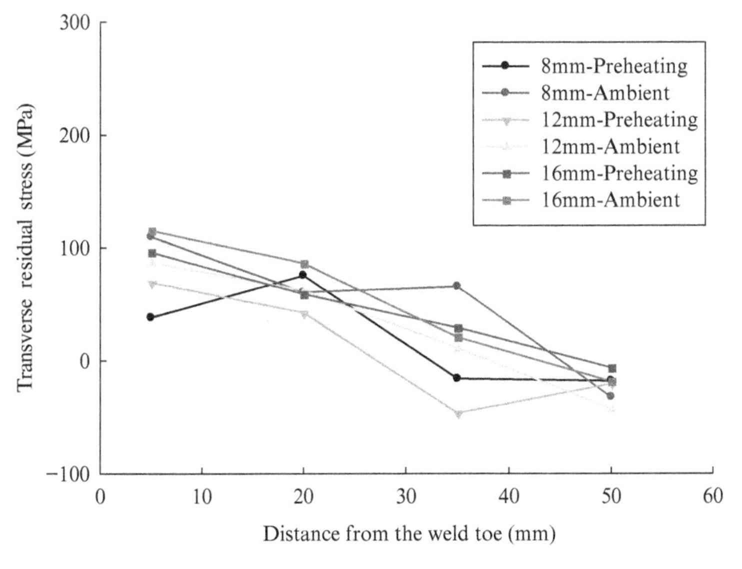

Figure 3.11 Residual stresses distribution along the transverse direction (*y* axis, gauges B, B1, B2, and B3) for 90° joints.

gauge B1, where the residual stress for the preheated joint is 24.2% higher than its ambient counterpart. At gauge B3 (50 mm from weld toe), it appears that the effect of preheating was not significant.

Figure 3.12 shows the residual stress distribution along the transverse direction (y axis) for 135° joints. Again, six curves are plotted to compare the preheating effect on residual stress distribution for different plate thicknesses. From Figure 3.12, it can be seen again that at 5 mm from the weld toe the maximum residual stress without preheating treatment exceeds 200 MPa (roughly 1/3 of yield strength) and preheating effectively relieved the magnitude of the residual stress (21.1%, 38.8%, and 37.8% reductions for 8 mm, 12 mm, and 16 mm joints, respectively). Similar to the 90° joints, at gauges B1 and B2 the relieving effect of preheating was reduced when compared with gauge B.

From Table 3.6 and Figures 3.11 and 3.12, it can be concluded that the magnitude of residual stress is highest near the weld toe (especially for the joints without preheating). Furthermore, in general, the residual stress magnitude decreases nonlinearly as the distance from the weld toe increases. When the distance exceeds 50 mm, the residual stress fluctuates around zero. Regarding the preheating effect on the distribution of residual stress, it was shown that preheating could effectively reduce the magnitude of the residual stress near the weld toe (within 5 mm) for both 90° and 135° joints. However, the stress relieving effects due to preheating reduce as the distances from the weld toe increases.

3.4.3 The Effects of Joint Angle

Figure 3.13 shows the residual stresses distributions along the transverse direction (y axis) for 90° and 135° joints welded at ambient temperature. From Figure 3.13, it can be seen that at gauge B (5 mm from weld toe), residual stress for 135° joints is higher than in that for 90° joints for all thickness (the differences are 40.9 MPa, 14.8

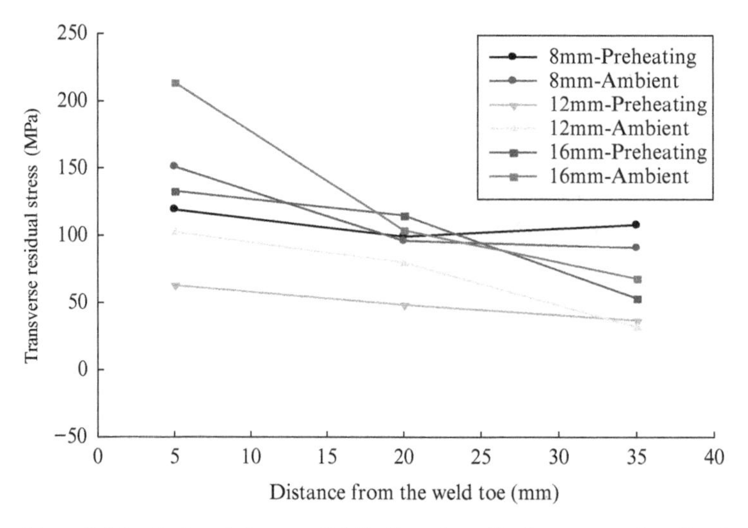

Figure 3.12 Residual stresses distribution along the transverse direction (y axis, gauges B, B1, and B2) for 135° joints.

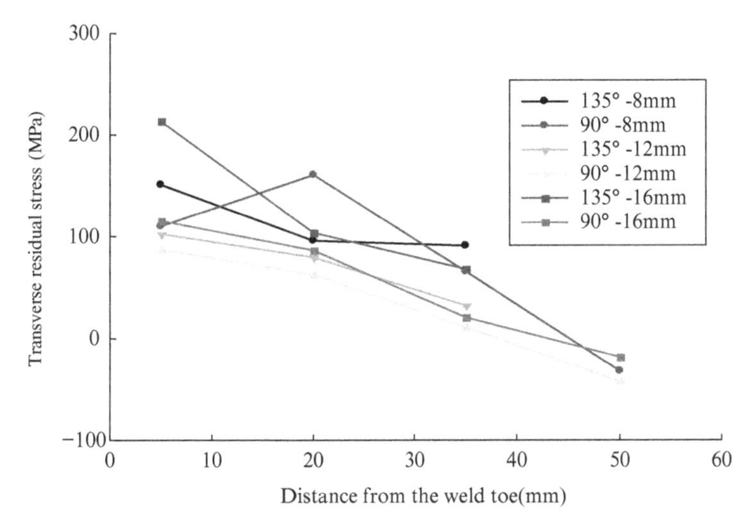

Figure 3.13 Residual stresses distribution along the transverse direction (*y* axis) for joints welded at ambient temperature.

MPa, and 98.1 MPa for 8 mm, 12 mm, and 16 mm joints, respectively). However, at gauge B1 (20 mm from weld toe) and gauge B2 (35 mm from weld toe), while the magnitude of residual stress for 135° joints is still higher than 90° joints, the effect of the welding angle is less significant.

Figure 3.14 shows the residual stresses distributions along the transverse direction (*y* axis) for 90° and 135° joints welded with preheating. Similar to the case of joints welded in ambient temperature, the residual stresses for 135° joints are higher than that for 90° joints, with the most noticeable difference at gauge B.

The effect of joint angle on the residual stress observed from the experimental study could be explained by considering the interrelationship and the influence of weld geometrical size and the difference of heat input (the energy transferred per unit length of weld). It can be seen in Figure 3.5 that the number of weld passes and, more importantly, the amount of welding material added to the joint required to form a 135° joint is more than that for a 90° joint. This obviously means that more heat energy is input to the 135° joint. In addition, a bigger weld also implies that a bigger molten pool and a bigger heat affected zone should be formed during the welding of the 135° joint. Hence, even when one assumes that the cooling rate for 90° and 135° joints is the same, higher residual stress is eventually generated in the 135° joints.

3.4.4 The Effects of Plate Thickness

From Figures 3.11 to 3.14 and Table 3.6, it can be seen that for both 135° and 90° joints, in general, higher residual stress was found for joints fabricated with 16 mm plates, regardless of whether the joint was preheated or not. However, it should be pointed out that at gauge B the residual stress does not always reduce as the plate thickness of the joint decreases. In fact, Table 3.6 indicates that in many cases (e.g. for

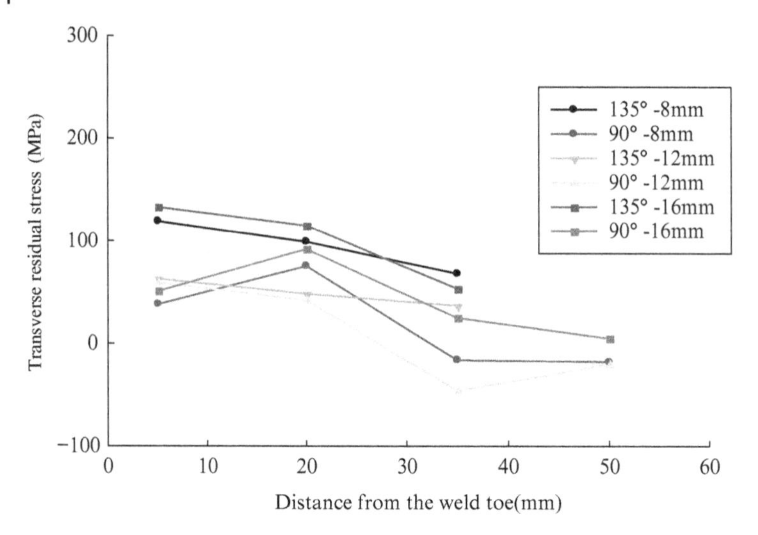

Figure 3.14 Residual stresses distribution along the transverse direction (y axis) for joints with preheating.

both 90° and 135° joints welded at ambient temperature), the residual stress at gauge B of a 8 mm joint could be higher than that of a 12 mm joint. This observation could be explained by the fact that for very thin (e.g., 8 mm) plates, while the temperature distribution generated by the welding process will be more uniform due to the small plate thickness, the joint will be cooled down more rapidly (by convection with surrounding air) due to its larger surface-area-to-volume ratio. As the thickness of the plate increases, while the temperature gradient will increase, the cooling rate will increase, too. Hence, it is possible that at certain thicknesses (e.g., 12 mm), the residual stress could attain a minimum as the reduction of the cooling rate could over-compensate for the increases in temperature gradient. Finally, as the plate thickness is further increased, the temperature gradient will be increased continuously while the cooling rate will become steady (mainly contributed by conduction through the plate) and thus the residual stress will be increased again.

Regarding the effect of preheating on the residual stress range for joints with different plate thickness, at gauge B for 90° joints welded at ambient temperature, the residual stress range for joints with difference thickness is 28.3 MPa (from 87.2 MPa for 12 mm joint to 115.5 MPa for 16 mm joint) while the corresponding range for preheated joints is 37 MPa (from 59.3 MPa for 12 mm joint to 96.3 MPa for 16 mm joint). For the case of 135° joints, the corresponding stress ranges for joints with and without preheating treatment are 70 MPa and 111 MPa, respectively. Hence, it can be seen that preheating seems to be more effective to reduce the effects of plate thickness for joints with larger intersection angles.

3.4.5 The Effects of Brace Plate Cutting

Figure 3.15 compares the effects of brace plate cutting on the residual stress distribution along the transverse direction (y axis of Figure 3.7) for 90° joints welded at

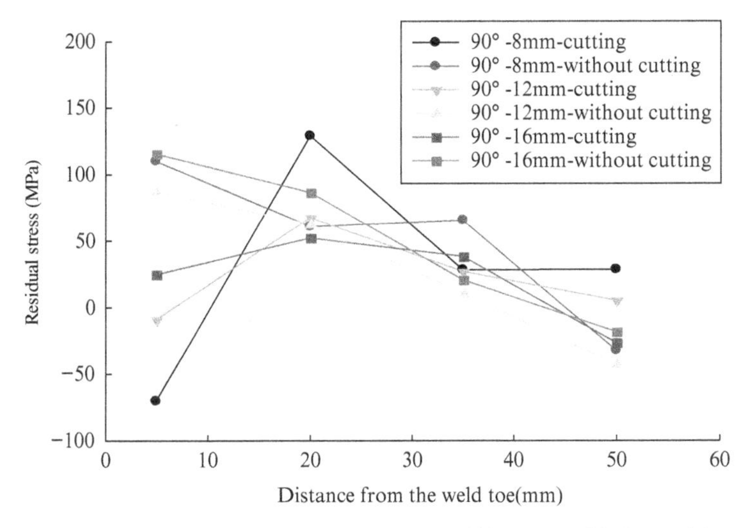

Figure 3.15 Effects of brace plate cutting for 90° joints welded at ambient temperature.

ambient temperature. From Figure 3.16, it can be seen that at gauge B (5 mm from weld toe), the magnitude of residual stress for joints without brace cutting is much higher than that with brace plate cutting. In fact, when the brace plate was cut off, compressive stress as high as 70.3 MPa was generated at gauge B. The differences in residual stresses at gauge B for 8 mm, 12 mm, and 16 mm joints with and without brace plate cutting are 180.4 MPa, 96 MPa, and 90.6 MPa, respectively. The corresponding differences are found to be reduced to 68.2 MPa, 5.0 MPa, and 34.1 MPa at gauge B1 (20 mm from weld toe). It is also observed that for the joints with brace plate cutting, the maximum stress value occurs at gauge B1 instead of gauge B. However, for the joints without brace plate cutting, the maximum stress value exists at gauge B for all different plate thicknesses.

Figure 3.16 compares the effects of brace plate cutting on the residual stress distribution along the transverse direction (y axis of Figure 3.7) for 90° joints welded with preheating treatment. From Figure 3.16, one can again observe that with brace plate cutting the magnitude of residual stresses at gauge B are much lower than the corresponding joints without brace plate cutting. In addition, the effect of brace plate cutting also reduces as the distance from the weld top increases.

Based on the results presented in Figures 3.15 and 3.16, it could be concluded that brace plate cutting could greatly disturb the residual stress distribution near the weld toe of the chord plate. Note that this finding agrees with works by Law et al. (2010), which also showed that the residual stress in the range of 15 mm from the cutting surface was relieved for butt welds in a plate. Even though the above conclusion was based on the results obtained from plate-to-plate T-joints with mechanical cutting close to the welding part, it indicates that any cutting close to the welding part should be avoid if one would like to conduct residual stress measurement near the weld toe.

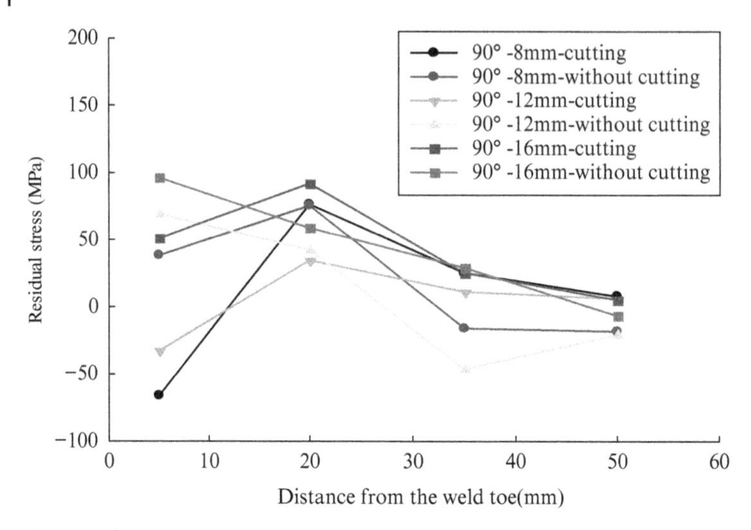

Figure 3.16 Effects of brace cutting for 90° joints with preheating.

3.5 Static Tensile Testing

3.5.1 Testing Rig

The Instron Model 8506 Dynamic Materials Testing System is introduced when tensile loading is to be applied in the specimen. With maximum tensile loading capability of 2000KN, it is an advanced multiprocessor-based control console which provides full digital control of a testing system. It consists of a closed load and four columns framed with a movable crosshead, a hydraulic actuator to apply a force, gripping mechanisms to hold the mechanical test specimen, and a load cell to measure the force. The position of the actuator, under closed loop control by controlling the hydraulic fluid flowing through a servo-valve supplying the actuator, is measured by a displacement transducer.

3.5.2 Strain Gauge Locations

After the measurement of residual stresses near the weld toe, all strain rosettes FRAS-2 are removed and another kind strain gauge, FLA-2 of TML strain gauge, which has only one grid, is adhered on the surface of the specimen. In the direction of width of the specimen, three strain gauges in a line are used and each one is close to a side of the drilled hole to dispel the stress concentration effect induced by the hole drilled in the residual stress testing (Figures 3.17–3.19). In the direction of length of the specimen, two lines strain gauges, which are 5 mm and 20 mm away from the weld toe respectively, are applied.

3.5.3 Testing Procedure

When the stress concentration factors and hot-spot stresses in the weld toe of a sample is focused, the stress distribution along direction of the weld toe is

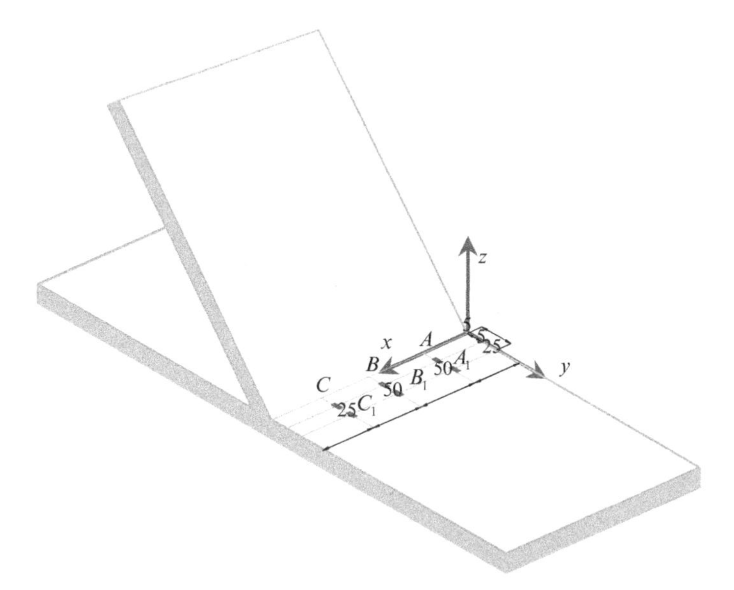

Figure 3.17 Strain gauges scheme for SCF measurement.

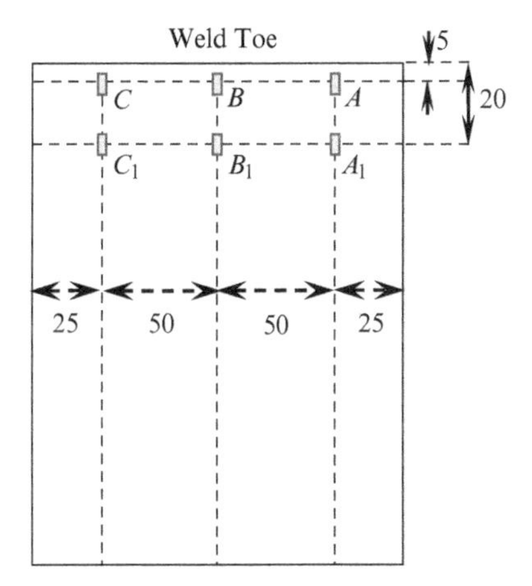

Figure 3.18 Plane view of scheme of strain gauges for static testing.

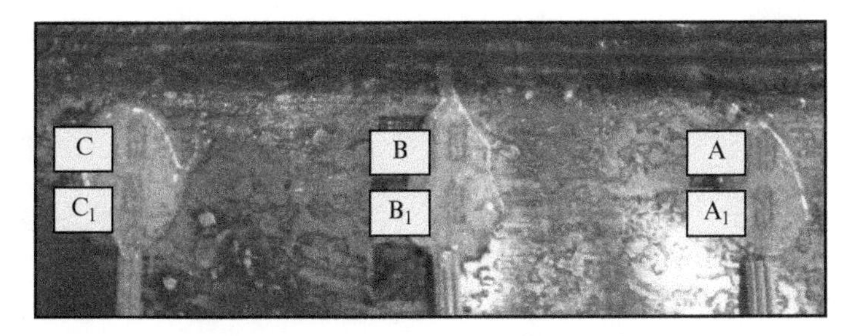

Figure 3.19 Strain gauge locations in specimen for static tensile testing.

investigated by monitoring the strains in the area when subjected to tensile loading in the brace plate, as shown in Figure 3.19.

To fix the specimen in the grip of the testing machine, a set of supporting joints, made of mild steel S355 with a thickness of 45 mm, was designed (Figure 3.20). It was designed in such a way that has nearly triple the thickness when compared with the specimens to make sure the failure will turn out in the specimens rather than the supporting joints. The specimen and supporting joints are connected by high-strength hexagon bolts of grade 10.9HR. Figure 3.21 is the full view of the testing machine when the specimen and the supporting set are fixed in the grips.

3.5.4 Testing Results

Table 3.7 gives a summary of SCF values for every case. The objective of static testing is to find the SCF values distribution in the weld toe when the joint is subjected to tensile loading. It is concluded that the most SCF values are located in the range from 1.0~2.0. Some values seem high since in several specimens notches exist near the weld toe. In most cases, the maximum SCF exists in the middle of the plate's width while, in some cases, this maximum value exists at one end of the plate's width.

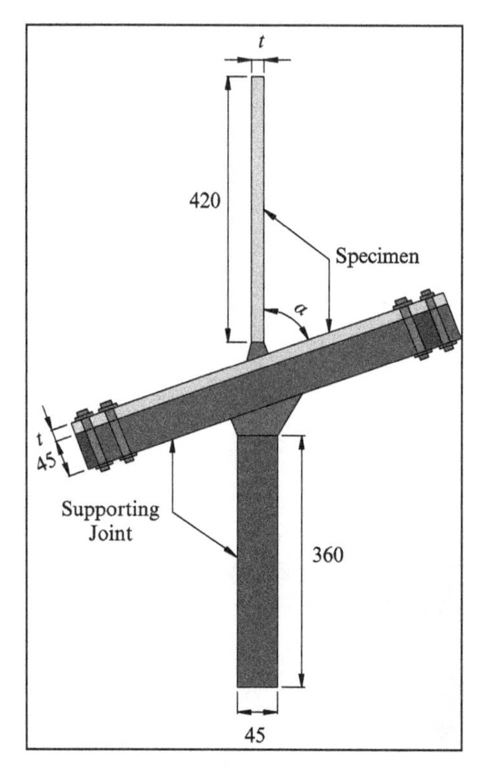

Figure 3.20 Assembly of the specimen and supporting joints.

Figure 3.21 Fixing of the specimen and the supporting joints in the testing machine.

Table 3.7 Summary of SCF values.

Joint geometry		Ambient temperature			Preheating		
Degree	Thickness (mm)	A	B	C	A	B	C
135	8	1.43	1.53	1.48	1.54	1.62	1.60
	12	1.61	1.82	1.74	1.48	1.53	1.30
	16	1.70	1.95	1.72	1.76	1.96	1.88
90	8	1.61	1.41	1.54	1.60	3.10	1.92
	12	0.8	1.20	1.63	1.77	2.38	2.05
	16	1.52	1.64	1.60	1.45	1.44	1.52

3.6 The Influence of Residual Stress on SCF Value

3.6.1 Analysis Method

In the experimental investigation, a carefully planned test was organized to analyze the residual stresses and SCF values in various HSS plate-to-plate joints. The residual stresses are compared in the view of welding condition, skewed angle, and plate thickness. Linear interpolation method was employed to estimate the residual stress at the weld toe.

In addition, linear interpolation method was used for hot-spot stress at the weld toe and SCF values without considering the residual stress that is shown. By comparing the SCF values without residual stress effect and another group of SCF values incorporating residual stress, the influence of residual stress on SCF is analyzed.

At any point in the specimen, the total stress, which is assumed to be smaller than the yielding stress σ_y, can be expressed as:

$$\sigma_{total} = \sigma_{rs} + \sigma_f \tag{3.5}$$

where, σ_{rs} is the residual stress at a certain point in the specimen and σ_f is the elastic stress caused by the applied load in the specimen. According to the definition of SCF, SCF_f can be written as:

$$SCF_f = \frac{\sigma_f}{\sigma_n} \tag{3.6}$$

where SCF_f is the stress concentration factor without considering residual stress and σ_n is the nominal stress caused by the applied load in the specimen.

When residual stress is considered, the *SCF* should be expressed as:

$$SCF_{total} = \frac{\sigma_{total}}{\sigma_n} = \frac{\sigma_{rs} + \sigma_f}{\sigma_n} = \frac{\sigma_{rs} + SCF_f \cdot \sigma_n}{\sigma_n} = \frac{\sigma_{rs}}{\sigma_n} + SCF_f \tag{3.7}$$

where SCF_{total} is the stress concentration factor including residual stress and stress caused by the applied load. By defining $SCF_{rs} = \frac{\sigma_{rs}}{\sigma_n}$, the equation can be written as:

$$SCF_{total} = SCF_{rs} + SCF_f \tag{3.8}$$

Therefore, the effect of residual stress can be measured by the residual stress factor (RSF), which is defined as:

$$RSF = \frac{SCF_{rs}}{SCF_{total}} \tag{3.9}$$

3.6.2 Results and Conclusions

Figure 3.22 shows the *RSF* values variations of 135° joints with different applied stress. When the applied stress is equal to 50 MPa at the 5 mm point, the RSF_s for the 8 mm, 12 mm, and 16 mm joints without preheating are equal to 0.65, 0.51, and 0.69, respectively. The corresponding values for the preheated joints with 8 mm, 12 mm, and 16 mm plate thickness are equal to 0.61, 0.41, and 0.58, respectively. Hence, it

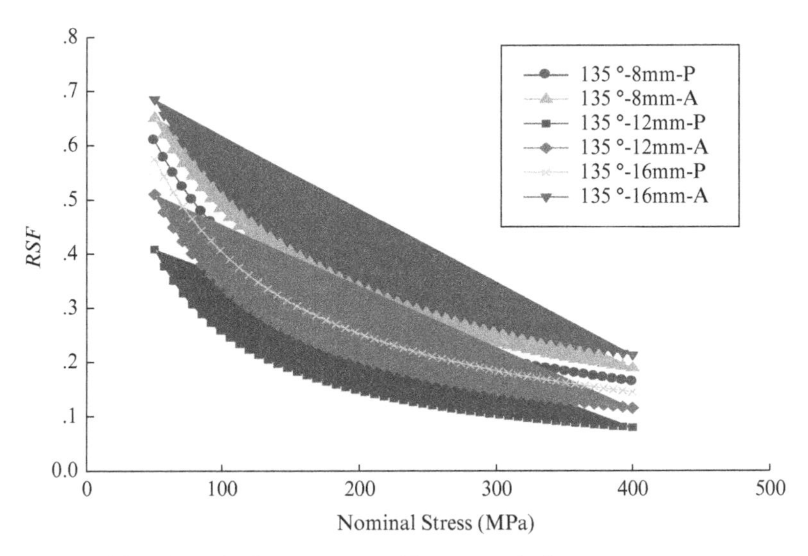

Figure 3.22 *RSF* of 135° joints under different nominal stresses.

can be seen that at the 5 mm point the *RSF* for the preheated joints was slight smaller than values without preheating. When the applied stress increases to 250 Mpa, *RSF* for the 8 mm, 12 mm, and 16 mm joints without preheating are equal to 0.27, 0.17, and 0.30, respectively. The corresponding *RSF* for the preheated joints with 8 mm, 12 mm, and 16 mm plate thickness are equal to 0.24, 0.12, and 0.20, respectively. Therefore, when the applied stress increases from 50 MPa to 250 MPa, the *RSF* for the 8 mm, 12 mm, and 16 mm joints without preheating were reduced by 58.5%, 66.7%, and 56.5%, respectively. For the 8 mm, 12 mm, and 16 mm joints with preheating, the corresponding *RSF* reduction is equal to 60.6%, 70.7%, and 65.5%, respectively. Thus, even though the residual stresses in joints without preheating are higher than the joints with preheating, the *RSF* reduction for the joints without preheating is faster than the preheated joint. When the applied stress increases from 250 MPa to 400 MPa, *RSF* reduces 30% approximately for all the specimens.

Figure 3.23 shows the relationship between *RSF* and the applied stress for 90° joints. For both preheated and ambient temperature joints, when the applied stress is 50 MPa the residual stresses in the weld toe are compressive so that the *RSF*s are negative and outside the range [0, 1]. For the joints with plate thicknesses of 8 mm, the *RSF*s at the 5 mm point are equal to -3.22 and -2.87 for ambient temperature and preheated joints, respectively. It is outside the scope [0, 1] and as a result the compressive residual stress at the 5 mm point is -105.2 MPa, which is larger than the applied stress at this moment. With further increase of the applied stress, the *RSF* approaches 0. Note that for a given applied stress, there exists a critical compressive residual stress which will lead to zero total stress (or $SCF_{total} = 0$) and therefore the *RSF* approaches infinity. Furthermore, when the applied stress is slightly smaller than this critical value, the *RSF* approaches positive infinity quickly while, when the applied stress is larger than this critical value, the *RSF* reduces quickly to minus

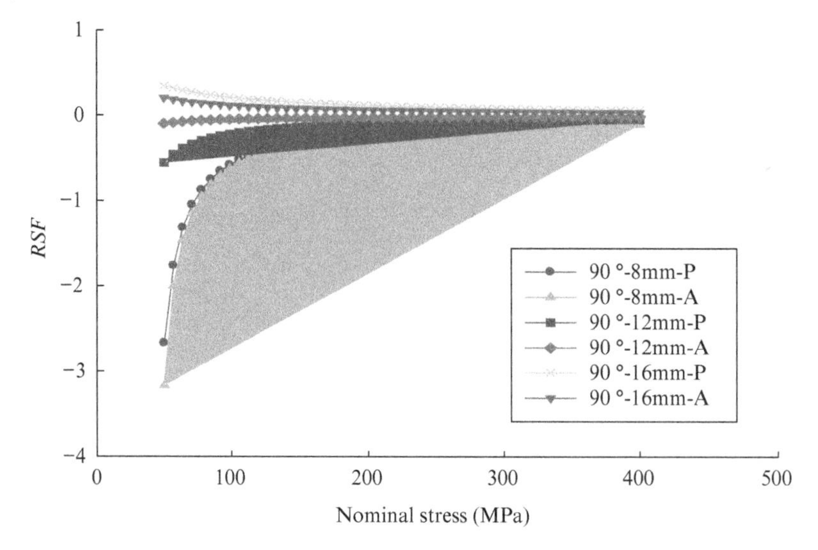

Figure 3.23 *RSF* of 90° joints under different nominal stresses.

infinity value. For the joint (90°, 12 mm, preheated), this critical value is roughly equal to 61.8 MPa. It also can be seen that when the applied stress is larger than 250 MPa, the *RSF* is comparatively small when comparing with when low stress is applied.

Based on the analysis in this section, some conclusions are drawn: when the applied stress ranges from 0 to 200 MPa, the SCF_{total} is quite high, and the effect of residual stress has significant influence on the SCF_{total}. *RSF* is more than 50%, so the residual stress strikingly changes the SCF_{total}. When the applied stress is higher than 200 MPa, the SCF_{total} goes to a stable value gradually. The influence of residual stress is less than the case when the applied stress is lower than 200 MPa. *RSF* reduces to 0~20% in these cases.

3.7 Conclusion and Summary

In this chapter, the results of a carefully designed experimental study conducted to investigate the residual stress distribution along the weld toe of high stress steel (HSS) plate-to-plate T/Y-joints were presented. Two sets of HSS plate-to-plate T/Y-joints with different geometrical details were fabricated with and without proper preheating treatment. The ASTM hole-drill method was then employed to measure the value of residual stress at critical locations of the joint near the weld toe. In addition, the effect of cutting the brace plate of the joint on the residual stress distribution was studied.

The experimental results indicated that while transverse residual stress with magnitude as high as one third of the yield strength of HSS could appear near the weld toe of the joint, proper preheating could significantly reduce the magnitude of the residual stress. In addition, it was found that the magnitude of residual stress increases as the plate thickness and the intersection angles of the joint increase. Furthermore, it was found that the magnitude of the residual stress reduces nonlinearly as the distance from the weld toe increases. By comparing the measurement results obtained from plate-to-plate T-joints with and without brace plate cutting, it was found that mechanical cutting operation near the welding part of a joint could considerably disturb the residual stress distributions along the weld toes. Hence, it is not advisable to perform any cutting prior to residual stress measurement.

4

Numerical Study of Residual Stress for High-Strength Steel Plate-to-Plate Joints

4.1 Introduction

Fusion welding, a process of melting the work pieces and adding filler materials to form molten pool, is frequently used in the construction of steel structures. In fusion welding, residual stress is induced due to the highly localized, non-uniform and transient heating, and the non-linearity of material properties under elevated temperature. The welding process may cause high tensile residual stress in the HAZ and lead to fatigue and fracture failures. Such effects are more obvious when the base metal is HSS, which shows lower capacity for strain hardening after yielding. In this case, large residual stresses may affect the fatigue and strength performances of the connection. Therefore, a good estimation of welding residual stress is necessary when HSS is used in building structural connections.

In the welding procedure, the melt metal cools and solidifies as a result of heat conduction in the metal and surface convection and radiation. Therefore, it is necessary to understand the temperature variation over time to evaluate the deformation and residual stress. The heat transfer process in the welding plays a key role in the formation of residual stress. It has the distinguishing feature of local concentration in heating. During the arc welding process, the structure is heated unevenly so that a high temperature gradient is induced in the area close to the fusion zone. At the same time, the microstructure of steel is changed in the melt zone. The expansion effect in the HAZ is limited by the nearby material so that compressive plastic strain is generated. Eventually, the melt shrinks with restriction of the close-by material in the cooling process so that the tensile residual stress is generated. It is a complicate coupled thermo-mechanical process and, consequently, numerical parametric studies are deemed to be important in order to understand the influences of different key welding parameters, such as the number of welding passes, the welding sequence, and the boundary constraint, on the distribution of residual stress.

Since testing is costly, time-consuming, and limited in obtaining the data, finite element modeling is broadly used in studying the residual stress formation and distribution caused by welding. Therefore, in this chapter, we examine a carefully selected sequentially coupled thermal-mechanical modeling procedure that was

Welded High Strength Steel Structures: Welding Effects and Fatigue Performance, First Edition. Jin Jiang.

developed for residual stress analysis for the HSS plate-to-plate joints that were studied in Chapter 3. Both 2D and 3D models are created to investigate the residual stress distribution in the HSS plate-to-plate joints, especially for the residual stress near the weld toe. Validation of the modeling will be conducted by comparing with the test results. After validating the accuracy of the modelling procedure, a small scale parametric study will be carried out to investigate the influence of some key welding parameters, such as the boundary condition, the preheating temperature, the number of welding passes, the welding speed, and the welding sequence on the magnitude and distribution of residual stress.

4.2 Modeling Procedure and Results for 2D Models

4.2.1 Overview

In this chapter, the finite element modeling package ABAQUS (2009) was used in the numerical modeling of the welding process. A sequentially coupled thermal-stress analysis was conducted by assuming that the stress solutions are dependent on the temperature fields while there is no inverse dependency. Sequentially coupled thermal-stress analysis was performed by first solving the non-linear transient thermal analysis heat transfer problem. The time-dependent temperature data was then fed into the stress analysis as a predefined field (Figure 4.1). During the thermal analysis, it was assumed that the stress due to the welding has negligible influence on the temperature field. In addition, the influence of phase change and microstructure evolvement was ignored in the modeling. However, the heat convection and radiation effect were both considered in the modeling. Table 4.1 gives a summary for the sequentially-coupled thermal-stress analysis procedure. The 2D plane strain models were created to obtain the residual stress variation of the cross section at the middle of the joint width (Figure 4.7, the cross section with the points B, B1, B2, and B3). The residual stress in the middle cross section of the plate width is generally higher than the residual stresses at both width ends, which is found in Chapter 3. The 2D finite element mesh used in the analysis is shown in Figure 4.2. Note that a larger element was used for the discretization of the base plate while a refined mesh was used for the weld filler and the connection.

4.2.2 Lumped Technique

To accurately model the multi-pass welding process, large computational costs are needed when fully 3D models are created. Therefore, a major concern for that case is to develop a reasonable modeling procedure that can save computational cost without sacrificing analysis accuracy. One popular approach to simplify the modeling procedure and to reduce the computational cost in welding modeling is the lumped technique (Free and Porter 1989; Rybicki et al. 1978; Rybicki and Stonesifer 1979; Ueda and Nakacha 1982). When the lumped technique is employed, two or more weld passes are condensed into weld blocks or lumps in the sequentially coupled analysis. With this procedure, rather than analyzing the temperature variation and stress formation for many weld passes, numerical simulations are only needed to be carried out for a few lumps.

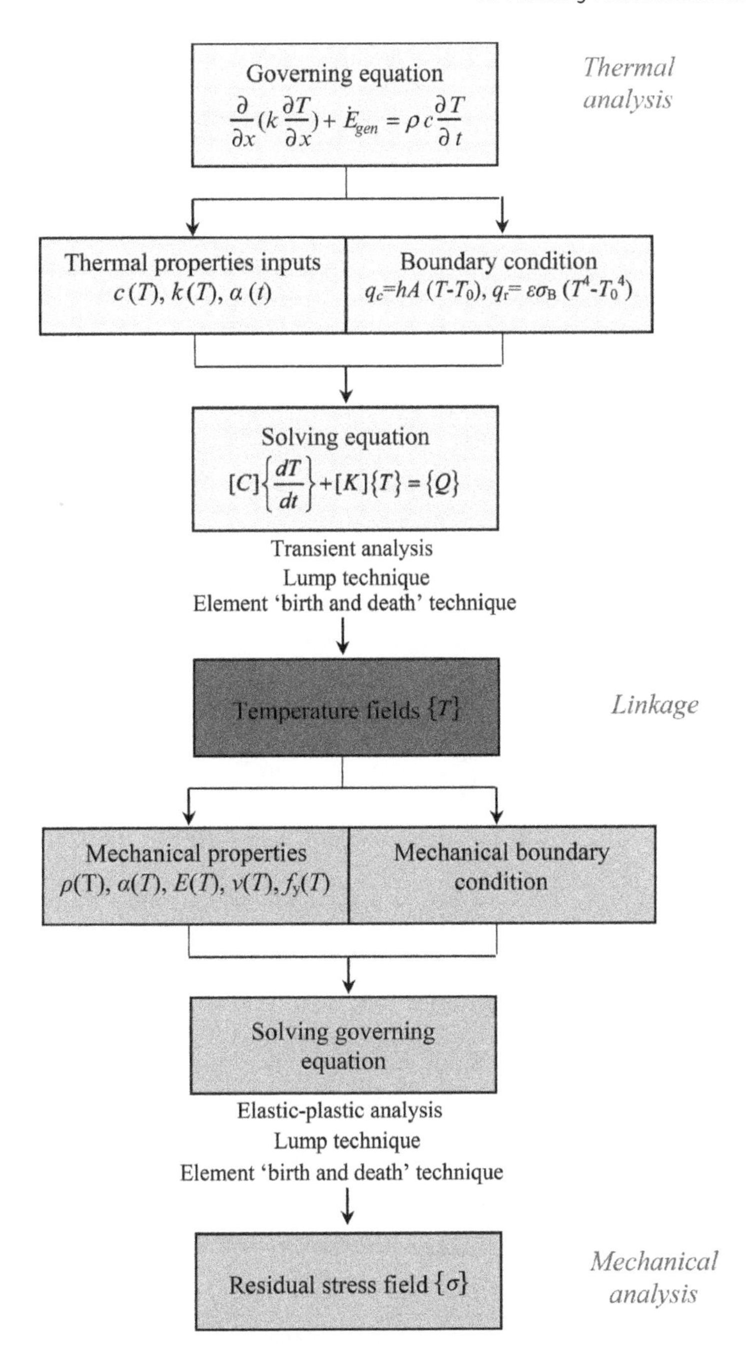

Figure 4.1 The flow chart of the modeling procedure.

By using different lumping schemes, the coarsest (but lowest computational cost) and the most accurate (but highest computational cost) solutions could be obtained by using just one lump and as many lumps as the actual weld passes, respectively. In practice, of course, a reasonable number of lumps should be

Table 4.1 Summary of the sequentially-coupled thermal-stress analysis procedure.

Item/Step	Heat transfer	Stress analysis
Element type	Four-node, linear-interpolation, heat- transfer element: DC2D4	Four-node bilinear plane strain quadrilateral, reduced integration element: CPE8R
Boundary Condition and Loading	Surface film Surface radiation	Temperature "load" via ODB file
Material Properties	Specific heat Density Conductivity	Elasticity Plasticity Thermal expansion
Numerical formulation	Transient thermal analysis	Elastic-plastic analysis

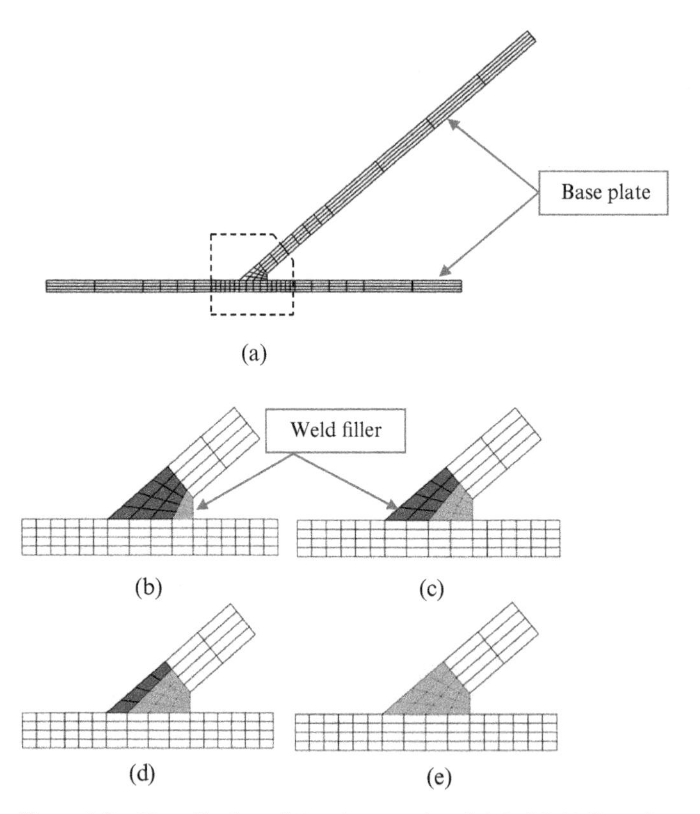

Figure 4.2 Discretization of the plate-to-plate joints (Light Gray elements: activated; Dark Gray elements: deactivated.)

selected in the modeling to balance the accuracy and the computational cost. In Chapter 3, it was recorded that 9 to 22 weld passes were employed in the welding of the plate-to-plate joints. They were lumped into four welding blocks in the numerical modeling.

4.2.3 Weld Filler Addition Technique

Three methods can be found to simulate the process of weld filler adding process (Figure 4.3). The first is named the full element method. In this method, both the base plate and weld pool are meshed together (Figure 4.3(a)). The second method, known as the element birth and death method (Figure 4.3(b)), is based on deactivating and reactivating the elements of the weld pool in the welding process. At the beginning, all the weld pool elements are deactivated. As the welding goes on, the corresponding elements are reactivated and included in the model. The third method is called the element movement method (Figure 4.3(c)). In this technique, the elements of the weld pool are separated from those of the base plate. A small distance d is shifted for the weld pool elements from the base plate. This way, the thermal and stress interaction between the base plate and weld filler is dependent on the distance d between them. Two-node elements are used to link the base plate elements and weld pool elements so that thermal and stress interaction can be transmitted.

The element birth and death technique is used in this study for the simulation of the addition of the weld filler materials (ABAQUS 2009). At the beginning of the modeling, all elements corresponding to the weld filler were deactivated. As the welding was proceeded on, the deactivated elements would be added into the model step by step. Such "birth and death" technique greatly simplifies the modeling procedure, as only a single finite element mesh is needed. However, this technique may introduce problems into the stress analysis since large displacements would be induced by the heating and cooling process, especially where many "death" elements are located. The boundary between old and newly-added elements may be strongly distorted. Furthermore, attempts to fit the fresh filler into the deformed model will cause stress redistribution of the residual stresses inherited from previous passes. To avoid the disadvantage of this technique, an alternative way is to include all elements in the model and to keep the elements corresponding to the weld not yet activated at a high softening temperature (Brickstad and Josefson 1998).

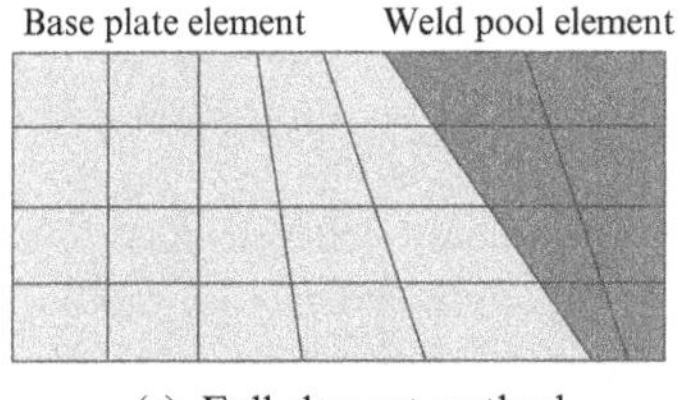

(a). Full element method

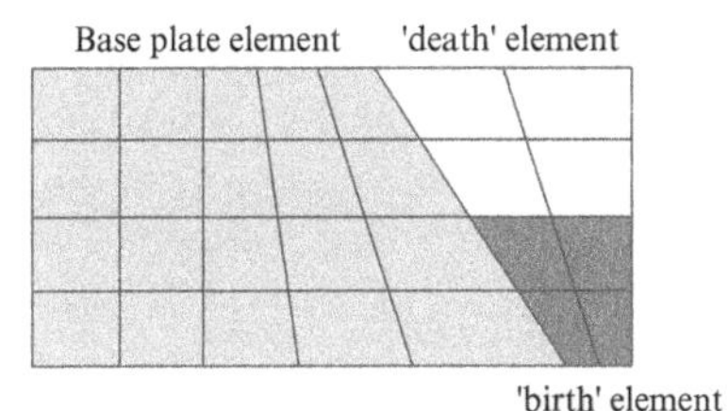

(b). Element birth and death method

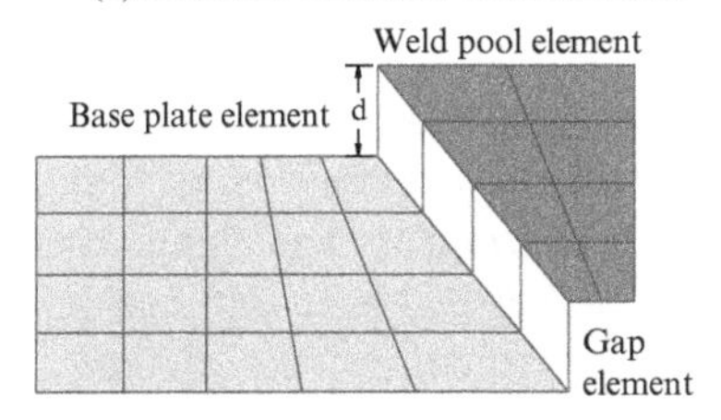

(c). Element move method

Figure 4.3 Modeling techniques for adding of weld filler.

In this chapter, to eliminate the adverse effects, a subsidiary step was performed after a new set of elements was added. In this step, a lower stiffness and yield stress were assumed for the newly added elements by multiplying a reduction factor of 0.35 to the true values. Each subsidiary step would last for 1.0s, and this time was excluded from the actual heat transfer and stress analyses since this step is used to transmit the deformed geometry from the previous pass to the next addition of weld filler. As shown in Figure 4.2(a), the model of the joint was first created by including all the elements for the base plates and weld filler. At the beginning of the analyses, all the elements of the weld filler were first deactivated. As the welding started, the first lump of weld filler was activated at 0.001s (Figure 4.2b). The second and the other lumps were then activated in sequence (Figure 4.2c–2e).

4.2.4 Heat Transfer Analysis

The heat transfer analysis was conducted based on the thermal energy balance principle, which could be stated mathematically as:

$$\int_{v} \delta_{\theta}\rho\dot{E}dV - \int_{v} \frac{d\delta_{\theta}}{dx} qdV = \int_{s} \delta_{\theta}\rho dS + \int_{v} \delta_{\theta} r dV \tag{4.1}$$

where q is heat flux per unit of current area crossing surface S from the environment into the body, r is heat flux per unit volume generated within the body, ρ is material mass density, $\dot{E}$ is internal energy per unit mass, and δ_{θ} is an arbitrary variation temperature field.

In the thermal analysis, transient non-linear analysis was conducted for 2D models to determine the temperature history throughout the whole heating and cooling process during the welding. The heat flux q was modeled by relating it with the welding speed v, the arc voltage U, and the arc current I. In the numerical model, a reasonable fraction of the heating duration Δt for the target cross section should be determined. The effect of moving the heat source is achieved by defining the amplitude of the heat source density curve. This is done by adjusting the heat source density amplitude curve until the melting temperature of 1300–1400 °C is attained in the molten zone and a maximum temperature of 500–600 °C is attained in the HAZ range. The distributed heat flux is defined by the Eq. 4.2 ($\eta = 0.8$ is the arc efficiency factor, l_w is the depth of each lump, h_w is the height of each lump, and v is the welding speed):

$$q = \frac{\eta UI}{l_w h_w v} \tag{4.2}$$

A typical heat transfer analysis for the welding process contains thermal material nonlinearities and change of heat dissipation surfaces. Thermal material nonlinearities include the thermal conductivity k and the specific heat c, which are functions of temperature (EC3 2005). Also, it should be noticed that the convection and radiation surfaces are always changed in the adding of weld filler procedure. With the adding of weld filler, heat dissipation due to convection and radiation were corresponding changed in the modeling. The convection coefficient h is defined as

15W/ m^2K and the emissivity ε is set as 0.2. In this study, the variations of thermal conductivity, specific heat and thermal expansion coefficients with temperature for the RQT701 HSS were obtained from Eurocode 3, Part 1–2 (Eurocode3 2005) and are shown in Figure 4.4.

In order to evaluate the effect of preheating, the number of weld lumps, the welding speed, and the welding sequence on the cooling rate of the joints, the average cooling rate at a given point of the joint was calculated. For a selected point of the joint, the average cooling rate K_t is defined as the temperature gradient between the time when the maximum temperature is attained to the time when the temperature at that point is dropped to 100 °C. During the modeling, it is considered fact that the thermal history, particularly soaking time at high temperatures and cooling time from 800 to 500 °C, determines the microstructure and mechanical properties for a given composition. However, at the same time, the cooling time from 400 to 150 °C is a controlling factor in the diffusion of hydrogen, the cold cracking of welds, and residual stress formation (Goldak et al. 1984). So, in order to give a comprehensive consideration, the cooling time from the maximum temperature to 100 °C was employed to calculate the cooling rate. The K_t can be computed by Eq. 4.3:

$$K_t = (T_{max} - 100°C)/(t_{max} - t_{100}) \tag{4.3}$$

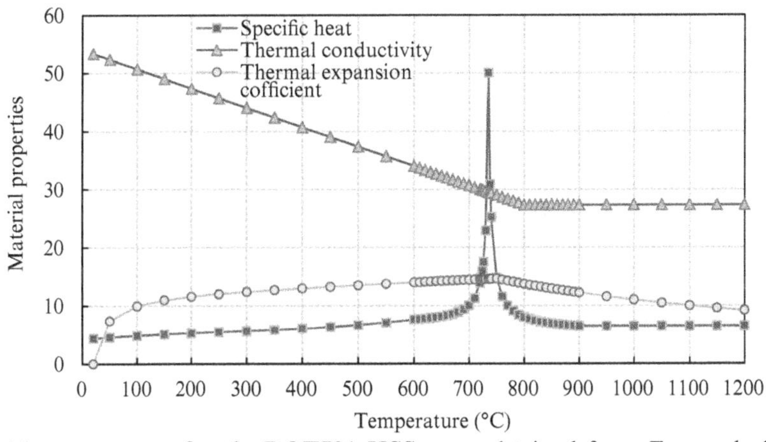

with temperature for the RQT701 HSS were obtained from Eurocode 3 Part 1-2

Symbol	Material properties	Unit
c	Specific heat	10^2J/(K.Kg)
k	Thermal conductivity	W/(K.m^2)
α	Thermal expansion coefficient	10^{-6}/K

(Eurocode3 2005) and are shown in Figure 4.4.

Figure 4.4 Thermal properties used in the modeling.

where T_{max} is the maximum temperature reached during the welding process, t_{max} is the time when the maximum temperature is attained, and t_{100} is the time when the temperature of the point is dropped back to 100 °C.

4.2.5 Mechanical Analysis

In the mechanical analysis, the temperature history obtained from the thermal analysis is input as a thermal loading in the structural model. To make the process of feeding temperature field to the stress analysis easy to handle, a compatible mesh with the same meshing topology and nodes numbering was used. However, it should be noticed that in order to obtain accurate stress analysis results, an 8-node bi-quadratic plane strain element with reduced integration, CPE8R, is applied for stress analysis, while a 4-node linear element DC2D4 was employed for the thermal analysis to attain stable results. Table 4.1 lists the detailed considerations for the modeling.

While the temperature-dependent thermal properties of the RQT701 HSS were obtained from Eurocode 3, Part 1–2 (Eurocode3 2005), the mechanical properties, including the Young's modulus E, the yield strength f_y, and the ultimate strength f_u were obtained by performing the coupon tests at normal and elevated temperatures as recommended by the relevant standards (BSI 1992). Figure 4.5 shows the

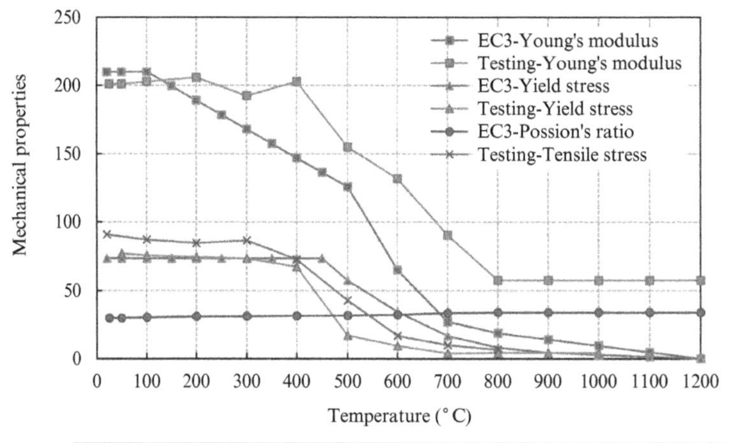

Symbol	Material properties	Unit
v	Passion's ratio	-
E-EC3	Young's modulus from EC3	GPa
E-Testing	Young's modulus from coupon test	GPa
f_y-EC3	Yield stress from EC3	(×10)MPa
f_y-Testing	Yield stress from coupon test	(×10)MPa
f_u-Testing	Tensile stress from coupon test	(×10)MPa

Figure 4.5 Mechanical properties from EC3 and testing.

variations of these mechanical properties as functions of temperature and their comparisons with the recommended values from Eurocode 3. Note that the Eurocode 3 is mainly applicable to the normal mild-structure steel and there are noticeable differences between the Eurocode 3 curves and the test curves, especially for the *E* values for the HSS plates used in this study.

4.2.6 Model Validation and Results

4.2.6.1 Model Validation

In this section, in order to validate the model accuracy, the numerical modeling results obtained from the sequentially coupled thermal-mechanical analysis were compared with the experimental data obtained in Chapter 3. Eighteen joints were fabricated for residual stress measurement. Therefore, eighteen numerical models were created in this part, corresponding to the joints tested in Chapter 3. Table 4.2 lists the detailed results at the selected points. By comparing with the test results, the numerical study seems able to accurately predict the residual stress distribution. The modeling procedures for these eighteen models are the same and two cases (135°, 12 mm, ambient and 135°, 12 mm, preheated) were selected herein to show the modeling results.

From Figure 4.6, for 90° joints with preheating, at the 5 mm point the stress differences between the modeling and testing are 31.0 MPa, 39.1 MPa, and 1.0 MPa for 8 mm, 12 mm, and 16 mm specimens respectively. At a selected point, the modeling results for three plate thicknesses are more convergent compared with experimental

Table 4.2 Modeling results at selected points (total: **18** models).

Specimen cases			Residual stress computed at the measuring points (MPa)						
θ (°)	t	Preheating	Weldtoe	5 mm	10 mm	15 mm	20 mm	35 mm	50 mm
90	8	Yes	96.7	69.2	42.3	41.1	40.5	40.1	21.7
		No	154.2	88.6	67.3	40.2	39.7	38.2	15.4
	12	Yes	167.8	98.4	77.3	42.1	40.3	39.6	5.7
		No	194.7	105.3	74.5	32.8	32.4	31.6	9.7
	16	Yes	197.8	97.3	81.5	41.1	40.8	40.3	12.7
		No	225.4	106.7	87.6	37.4	37.1	36.5	15.8
135	8	Yes	211.1	102.4	74.3	44.9	44.1	43.6	14.9
		No	254.6	126.7	65.4	42.8	42.2	41.7	17.6
	12	Yes	224.1	101.4	47.8	38.8	35.3	34	21.7
		No	293.3	86.3	36.5	31.5	27.9	32.6	22.9
	16	Yes	234.1	121.7	54.3	51.3	50.7	50.1	25.7
		No	278.5	149.4	57.9	53.4	52.7	52.3	23.1

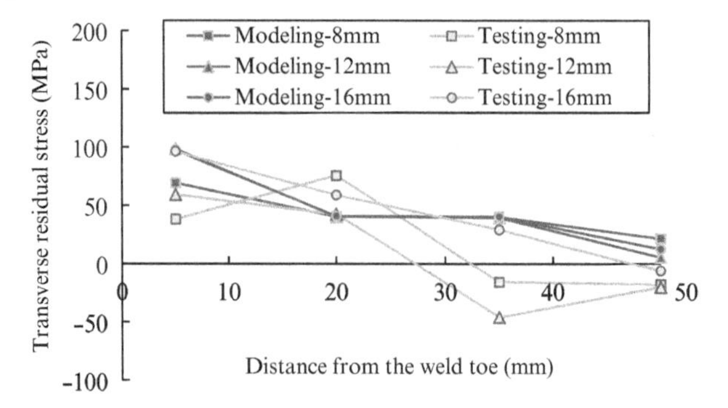

Figure 4.6 Comparison of modeling and testing results for $\theta = 90°$ joints with preheating.

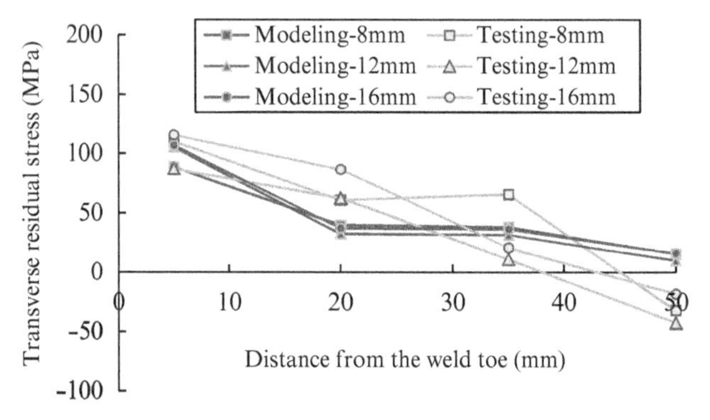

Figure 4.7 Comparison of modeling and testing results for $\theta = 90°$ joints welded at ambient temperature.

results. This phenomenon may be caused by the actual operation of the SMAW, such as breaking-off, insufficient weld, geometrical error, and so on. However, these stress differences between the modeling and testing are smaller than 50 MPa at all the positions. Similarly, as shown in Figure 4.7, the modeling results coincide well with testing results for 90° joints welding at ambient temperature, especially at the 5 mm point.

Figures 4.8 and 4.9 give another two residual stress comparisons between the modeling and testing for 135° joints. For the joints (135°, preheating), the difference of the transverse residual stress between the modeling results and the testing results are 16.7 MPa, 38.6 MPa, and 11.1 MPa for 8 mm, 12 mm, and 16 mm specimens, respectively, at the 5 mm point. For the joints (135°, ambient temperature), at the 5 mm point, modeling quality seems to be acceptable except the joint with 16 mm thickness, whose stress difference is 64.2 MPa. Variations during the welding would possibly cause this exception.

4.2.6.2 Numerical Modeling Results

4.2.6.2.1 Heat Transfer Results

To understand the temperature variations and residual stresses formation process during the welding, the temperature distributions near the weld region are studied

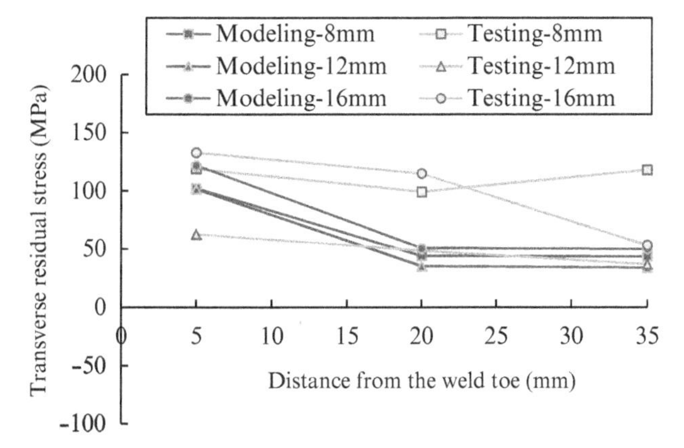

Figure 4.8 Comparison of modeling and testing results for $\theta = 135°$ joints with preheating.

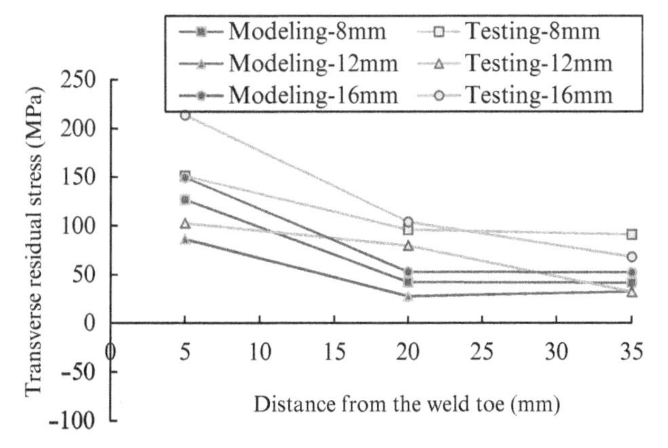

Figure 4.9 Comparison of modeling and testing results for $\theta = 135°$ joints welded at ambient temperature.

first. Figure 4.10 shows the temperature history near the welding region (area enclosed by the dash lines in Figure 4.2(a) for the joint of 135°, 12 mm, welded at ambient temperature). Figure 4.10(a) shows the temperature distribution for the joint at 1.0s after the welding started. At that moment, as the 2nd, 3rd, and 4th lumps of the weld were still deactivated, heat energy was mainly propagated only to a small localized area near the chord and the brace plate intersection. Figures 4.10(b), 4.10(c), and 4.10(d) show the temperature distributions at 1 second after the 2nd, 3rd, and 4th lumps were added, respectively. From Figures 4.10(a) to 4.10(d), it can be seen that as the welding proceeded on, the heat energy transferred from the center of the heat source (the activated lumps) to the ends of the plates. When the propagation time was 300s (a transient moment in the cooling period), the weld and the regions in the chord plate which were directly below the weld attained the highest temperature compared with other parts of the joint. When the propagation time was 2500s, shown in Figure 4.10 (f), the steady state temperature distribution was reached for the whole joint and the final residual stress distribution for this stage was formed.

4.2.6.2.2 Residual Stress Results

Figure 4.11(a) and 4.11(b) show the residual stress at the final steady stage. Figures 4.12 and 4.13 show the transverse residual stress for the joints welded at ambient temperature and 100 °C preheating, respectively, at different propagation times. When the propagation time was 57.7s (the moment when the 2nd lump is about to

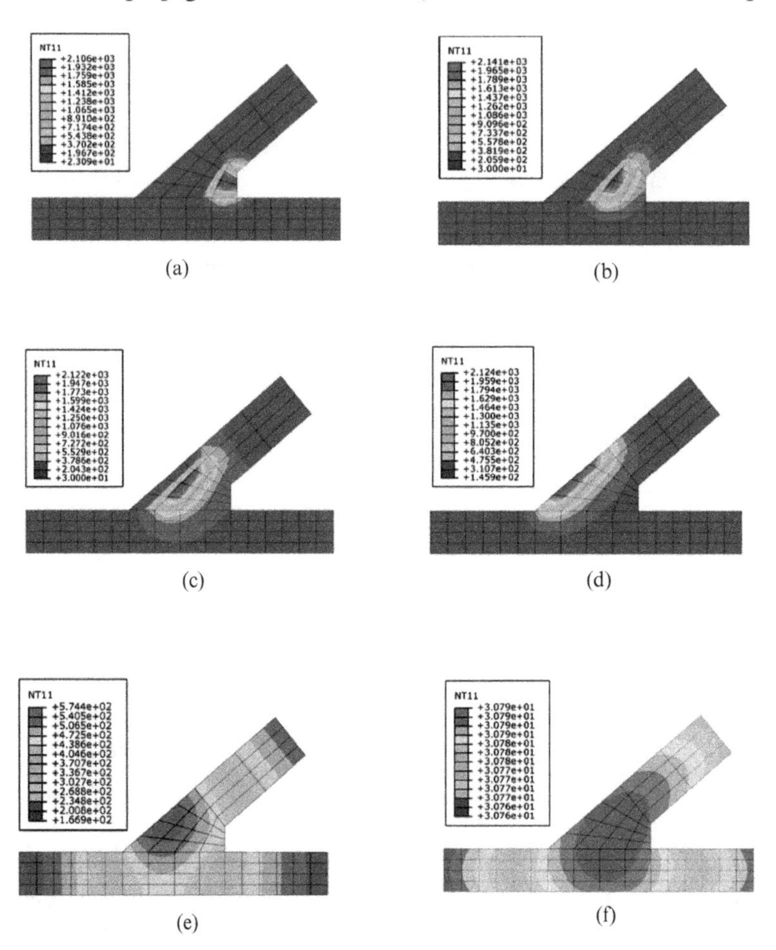

Figure 4.10 The temperature distributions at different times ($\theta = 135°$, $t_1 = 12$ mm, welded at ambient temperature) (a) t = 1.0s, (b) t = 58.7s, (c) t = 117.4s, (d) t = 176.1s, (e) t = 300s, (f) t = 2500s.

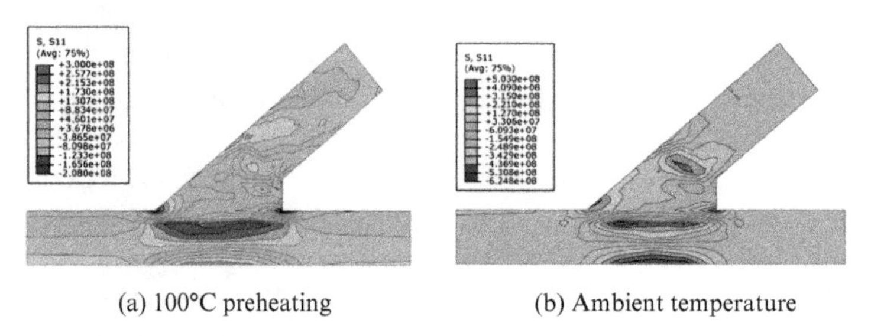

(a) 100°C preheating (b) Ambient temperature

Figure 4.11 The residual stress near weld in joints ($\theta = 135°$, $t_1 = 12$ mm).

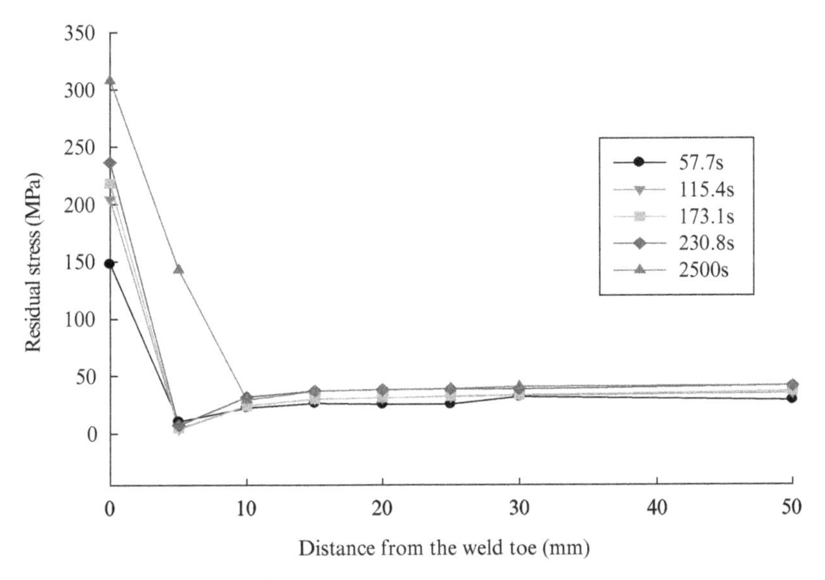

Figure 4.12 Relationship between transverse residual stress and distance from the weld toe ($\theta = 135°$, t_1 = 12 mm, preheated).

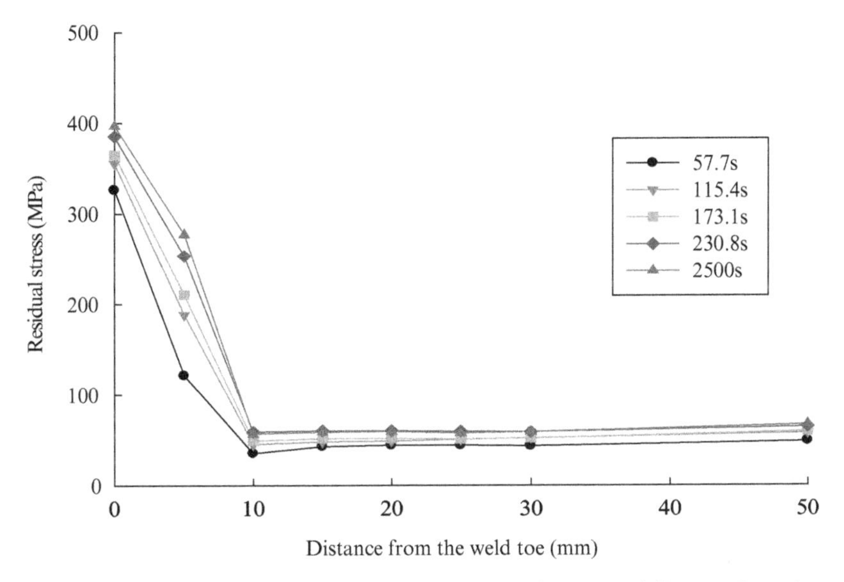

Figure 4.13 Relationship between transverse residual stress and distance from the weld toe ($\theta = 135°$, t_1 = 12 mm, ambient temperature).

be added), the transverse residual stress at the weld toe was increased to 330 MPa for the joint welded at ambient temperature. Comparatively, the transverse residual stress at the weld toe is 148.6 MPa for the joint welded with 100 °C preheating. Therefore, preheating can significant reduce the magnitude of the residual stress during the heat propagation process before the 2nd lump of weld is added. Note that at the final state (t=2500s), the transverse residual stresses at the weld toe are 408.7

MPa and 316.5 Mpa, respectively, for joints welded at ambient temperature and with 100 °C preheating.

For the joint welded at ambient temperature, the residual stress is mainly formed during the time interval between the beginning of the welding process to the moment when the 2nd lump of weld was added. Comparatively, only a small increase of the magnitude of the residual stress happened during the remaining welding process. For the joint welded with preheating, the residual stress components that formed both in the first state (0–57.7s) and the final cooling state (230.8–2500s) are both significant. Therefore, for the multi-pass welding, the final cooling process should be handled more carefully in the welding operation than when the joint is welded at ambient temperature.

It is also shown in both Figures 4.12 and 4.13 that the transverse residual stress in the region within 10 mm from the weld toe is significantly higher than the other parts. However, when the distance from the weld toe is further than 10 mm, the magnitude of residual stress seems stable and does not change much.

Figure 4.14 shows the average cooling rate for the two joints (135°, 12 mm, ambient and 135°, 12 mm, preheating). It can be seen that when the distance from the weld toe is less than 15 mm, the average cooling rate in the preheating joint is smaller than the ambient one. The preheating can effectively reduce the average cooling rate for the joint.

4.3 Modelling Procedure and Results for 3D Models

4.3.1 Overview

In order to obtain a more in-depth understanding of the residual stress distribution along joint width direction, 3D models corresponding to two joints ($\theta = 135°$, $t_1 = 12$ mm, *ambient temperature and* $\theta = 135°$, $t_1 = 12$ mm, *preheating*) were created. Similar to the 2D modeling procedure, the sequentially thermal-mechanical

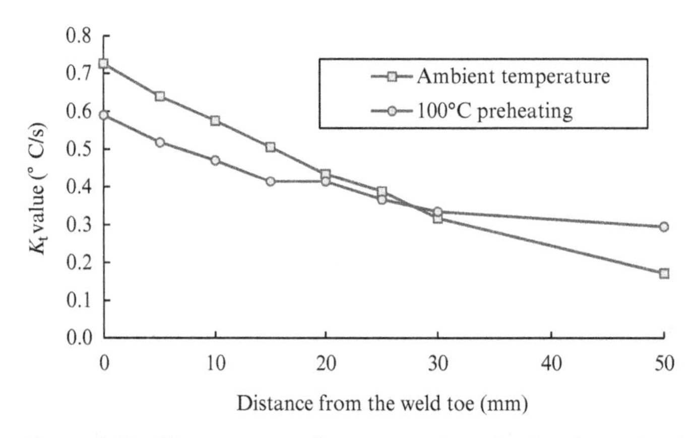

Figure 4.14 The average cooling rate at selected points for preheating and ambient temperature cases ($\theta = 135°$, $t_1 = 12$ mm).

coupled analysis method is used for the 3D models. In the SMAW and FCAW processe, rather than moving the arc to the starting point of a weld pass, the new pass of weld filler is often added in a continuous manner. For the studied HSS plate-to-plate joints, the new pass starts from the end position of the previous weld pass and goes along the reverse direction to the previous weld pass. Unlike the 2D models, which only consider the residual stress at the middle of the joint width and ignore the weld filler adding path in the joint width direction, 3D models can effectively evaluate the influence of weld filler adding direction on the final residual stress distribution.

4.3.2 Heat Source Model in 3D Analysis

An accurate analysis of thermal cycles is required to obtain accurate predictions of residual stress in 3D modeling. This can be traced back to the late 1930s, when Rosenthal (1946) first applied the Fourier law to moving heat sources. However, the main shortcoming of this solution is the misfit of the temperature field near the fusion and heat-affected zones. To accurately capture the temperature near the arc, Pavelic et al. (1969) mentioned that the heat source should be distributed, and proposed a Gaussian distribution of the heat flux deposited on the surface of the work piece. Considering that the temperature gradient in front of the heat source is different from in the rear, a double ellipsoidal model of power density distribution was mentioned by Goldak (Goldak and Akhlaghi 2005). In the double ellipsoidal model (Figure 4.15), the front half of the source is the quadrant of one ellipsoidal source and the rear half is the quadrant of another ellipsoid. The fractions f_f and f_r of the heat deposited in the front and rear quadrants are needed, where $f_f + f_r = 2$. It is recommended by Goldak that f_f and f_r can use 0.6 and 1.4, respectively. The power density distributions inside the front quadrant and the rear quadrant can be expressed as:

$$\begin{cases} q_1(x,y,z)=\dfrac{6\sqrt{3}f_fQ}{a\cdot b\cdot c_1\pi\sqrt{\pi}}e^{\frac{-3x^2}{a^2}}e^{\frac{-3y^2}{b^2}}e^{\frac{-3z^2}{c_1{}^2}} \\ q_2(x,y,z)=\dfrac{6\sqrt{3}f_fQ}{a\cdot b\cdot c_2\pi\sqrt{\pi}}e^{\frac{-3x^2}{a^2}}e^{\frac{-3y^2}{b^2}}e^{\frac{-3z^2}{c_2{}^2}} \end{cases} \tag{4.4}$$

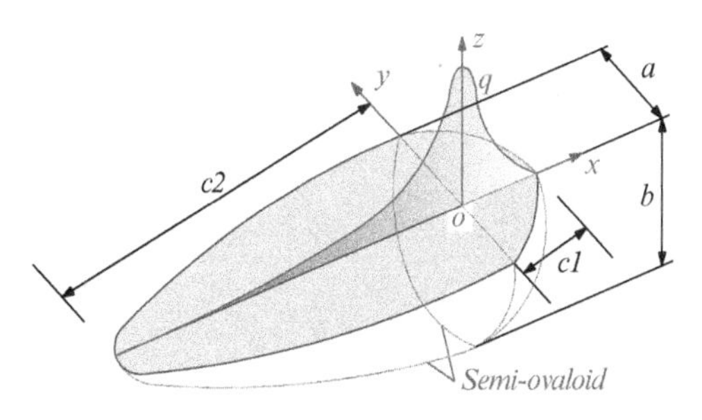

Figure 4.15 Double ellipsoidal heat source model.

where $Q = \eta \cdot U \cdot I, \eta$ is heat source efficiency, U is voltage of electric arc, I is current of electric arc, and a, b, c_1, c_2 are ellipsoidal parameters.

In this analysis, the double ellipsoidal model is used to predict the thermal and stress fields during the welding. Since there are four weld clusters at the corners of the box and circumference-going weld filler added at the intersection of the joint, a FORTRAN program is developed to describe the moving of the heat source (Appendix 2 gives the source code).

4.3.3 Modeling for the Weld Filler Adding Process

Figure 4.16 shows the actual welding sequence and direction for the 3D HSS plate-to-plate joint in the fabrication procedure. Since the lumping technique is used to reduce the computational cost in the modeling procedure, the modeling for welding sequence and direction needs to be simplified correspondingly. In this study, the weld is divided into two clusters. For each cluster, the welding sequence and direction are shown in Figure 4.17. Considering the fact that weld passes are not

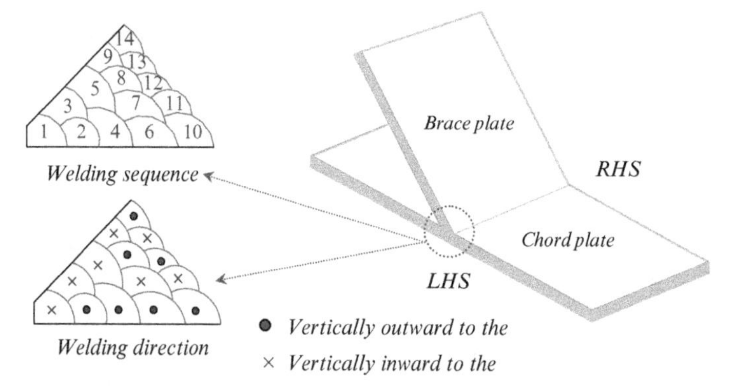

Figure 4.16 Welding direction in the selected HSS plate-to-plate joint ($\theta = 135°$, $t_1 = 12\,mm$, ambient temperature and $\theta = 135°$, $t_1 = 12\,mm$, preheating, *LHS:* left hand side, *RHS:* right hand side).

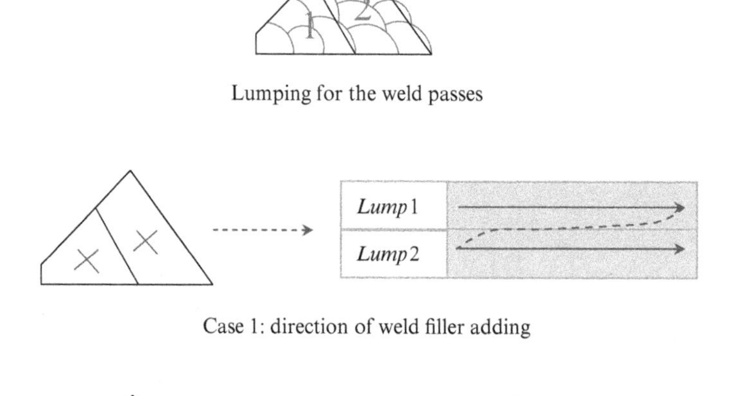

Figure 4.17 Modeling of welding direction along joint width.

added in the same direction along the joint width, as shown in Figure 4.16, the impact of welding direction is studied in this section. To accurately obtain residual stress results, the most accurate modeling procedure is to exactly follow the multi-pass welding procedure. In this study, two cases (one corresponding to a joint in which all weld passes start from the left end, the other one corresponding to a joint in which the first lump starts from the left end and the second lump starts from the right end, Figure 4.17), are studied.

The meshing and the welding filler adding process is shown in Figure 4.18. The element birth and death technique is used to simulate the weld filler adding process. In the operation, the weld filler elements are divided into 5 sets for each weld lump, which are reactivated at corresponding analysis steps. So, the weld filler is reactivated at 10 analysis steps. The convection and radiation surfaces are renewed for each step because when a new set of weld filler is added, the contact surface with air is renewed.

Figure 4.19(a) shows the start moment of activating weld filler elements. At this moment, several elements corresponding to the weld filler are activated and the heat source starts to move along the arc torch travelling path. Figure 4.19(b) shows the activated elements in the weld filler at the moment when the 1st lump of weld is fully achieved. The center heat source at this moment moves to the other end of the chord plate width. From the temperature contour, it can be observed that when the distance from the heat source center goes further, the temperature drops quickly. Figure 4.19(c) shows the moment when the elements of the 2nd lump of weld are to be activated. Figure 4.19(d) shows the case when elements at the weld filler are fully activated.

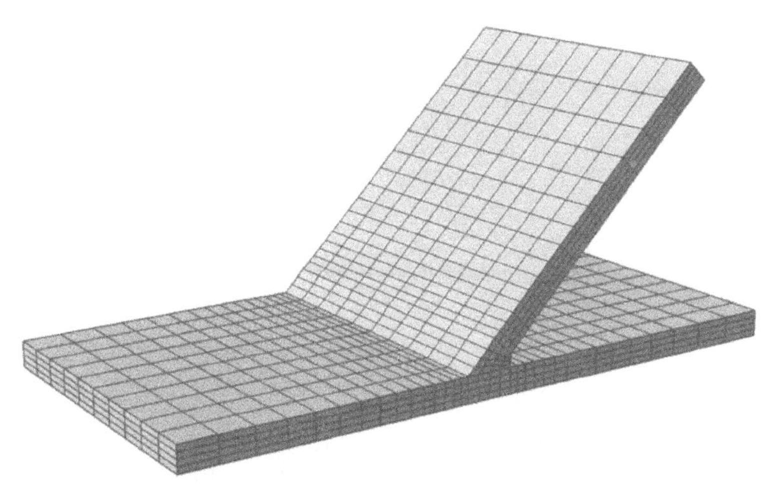

Figure 4.18 Meshing for the HSS plate-to-plate joint.

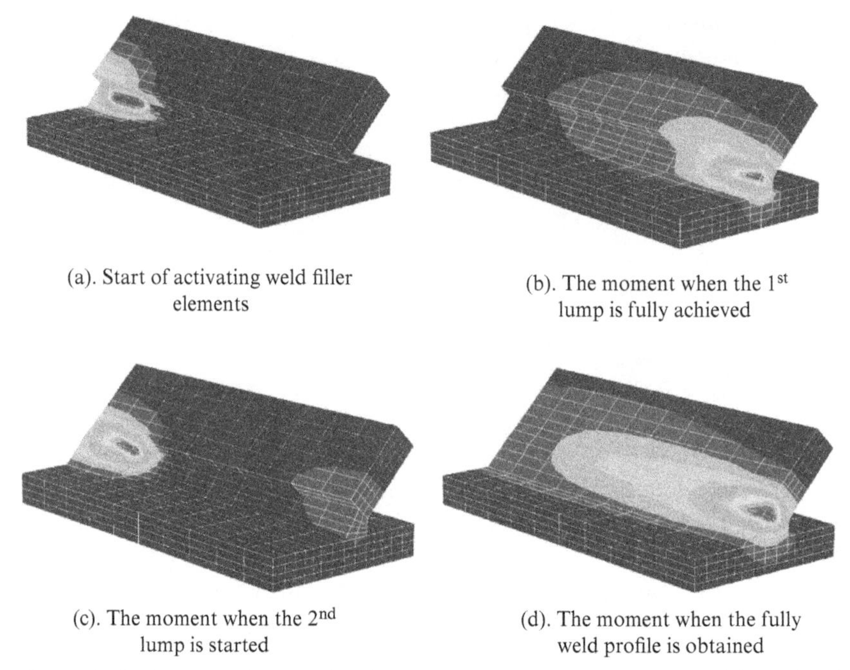

(a). Start of activating weld filler elements
(b). The moment when the 1st lump is fully achieved
(c). The moment when the 2nd lump is started
(d). The moment when the fully weld profile is obtained

Figure 4.19 Element birth and death technique used in the modeling (welding direction, *case 1*).

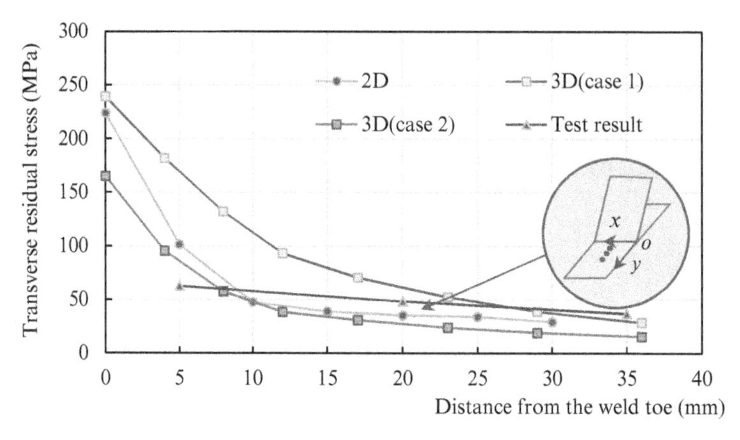

Figure 4.20 Comparison for 2D and 3D modeling and test results ($\theta = 135°$, $t_1 = 12$ mm, ambient temperature).

4.3.4 Modeling Validation

Figures 4.20 and 4.21 show the transverse residual stress predicted by 2D and 3D models and the test results ($\theta = 135°$, $t_1 = 12$ mm) obtained from Chapter 3. For the transverse residual stress at the chord weld toe, the 2D model result agrees well with two 3D models. In addition, all of these modeling results agree well with test results.

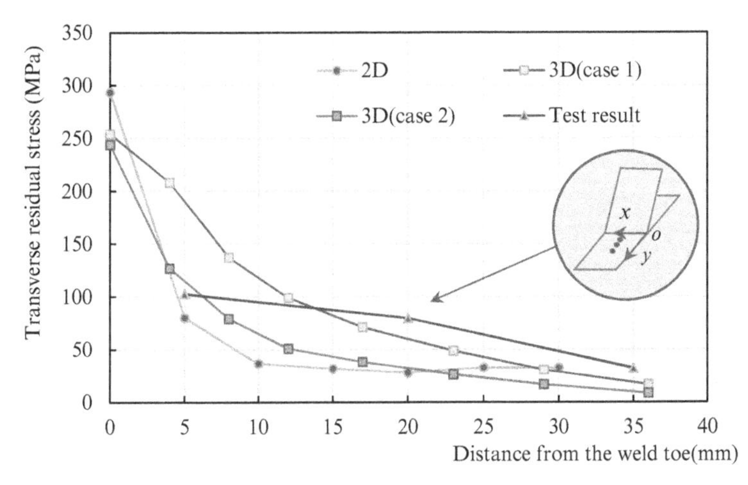

Figure 4.21 Comparison for 2D and 3D modeling and test results ($\theta = 135°$, $t_1 = 12$ mm, preheating).

4.3.5 Modeling Results

4.3.5.1 Ambient Temperature Joint

The final state of the residual stress field when the welding direction is case 1 (Figure 4.17) is present in this section. For the final results when the welding direction is case 2, details can be found in Appendix 1.

Figure 4.22(a) shows the transverse residual stress distribution in the joint ($\theta = 135°$, $t_1 = 12$ mm, ambient temperature, welding direction: case 1) when it is fully cooled down to room temperature. The transverse residual stress is highly localized near the weld and when the distance from the weld is beyond 50 mm, the magnitude of the transverse residual stress is near zero, which agrees with the observation in the testing. It can be observed that at both chord width ends, negative stresses (compressive residual stress) exist at the top surface of the chord plate. High tensile residual stress turns out in the middle of the chord width.

Figures 4.22(b) to 4.22(e) gives the transverse residual stress distributions at selected cross sections. Figure 4.22(b) shows the transverse residual stress at the cross section obtained by cutting the joint with surface *a* as shown in Figure 4.22(a). At this cross section, the maximum compressive residual stress, which is located on the top and bottom surfaces of the chord plate, turns out near the midpoint along the length direction. For the residual stress variation in the width direction, it is shown that high tensile residual stress exists at the midpoint along width direction while compressive residual stress turns out at both ends. For the residual stress variation in the depth direction, a layered distribution of residual stress can be found at this cross section in depth. The maximum tensile residual stress turns out on the top surface of the chord plate. With increase an in depth at this cross section, the residual stress decreases. Figures 4.22 (c) and 4.22(e) show similar distribution of transverse residual stress in the localized part under the weld. Different from the two ends, the part near the midpoint under the weld shows high tensile residual stress on the top surface of the chord plate.

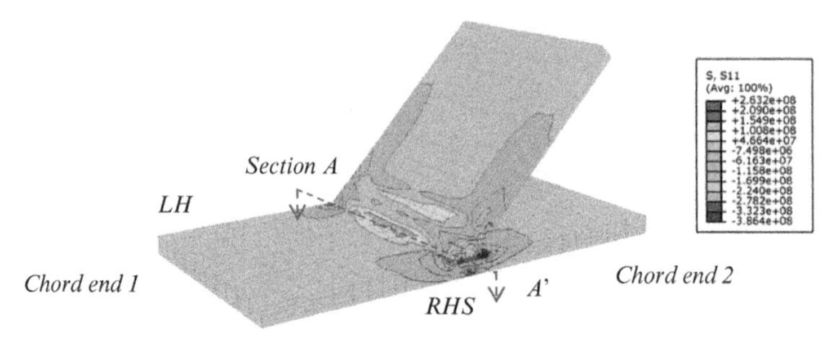

(a). The transverse residual stress at the final state (cooled down)

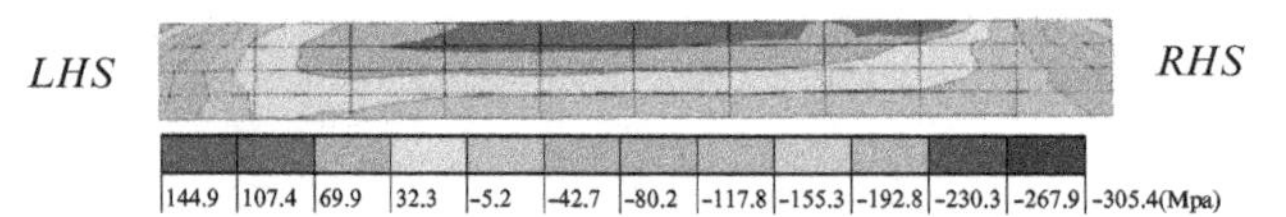

(b). The transverse residual stress at the weld toe (cut by section A-A')

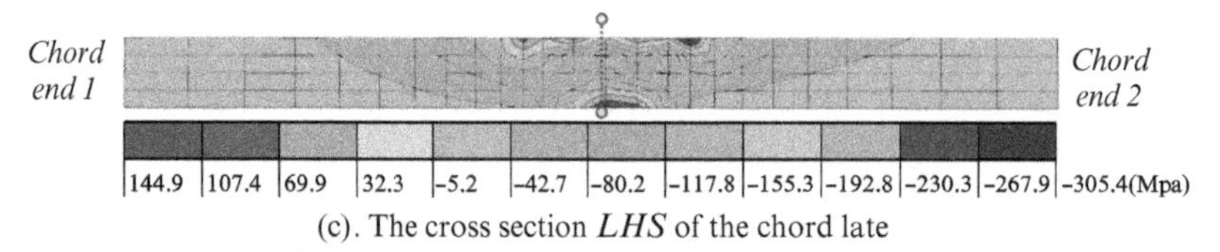

(c). The cross section *LHS* of the chord late

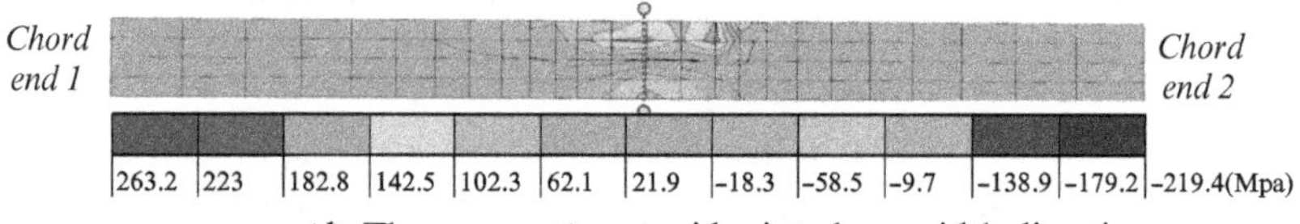

(d). The cross section at midpoint along width direction

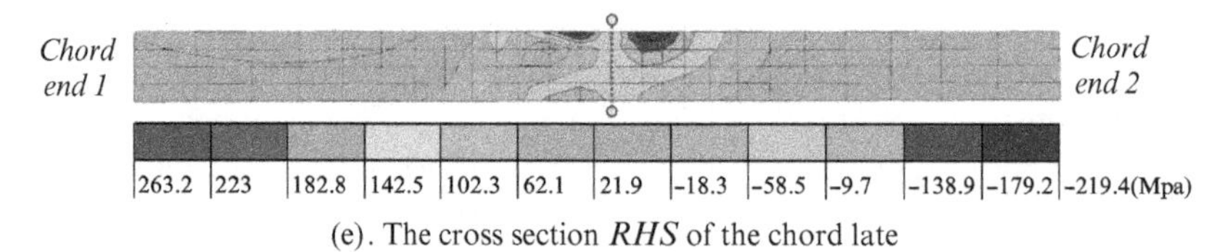

(e). The cross section *RHS* of the chord late

Figure 4.22 The transverse residual stress profile for the ambient temperature specimen (*LHS:* left-hand side, *RHS:* right-hand side)

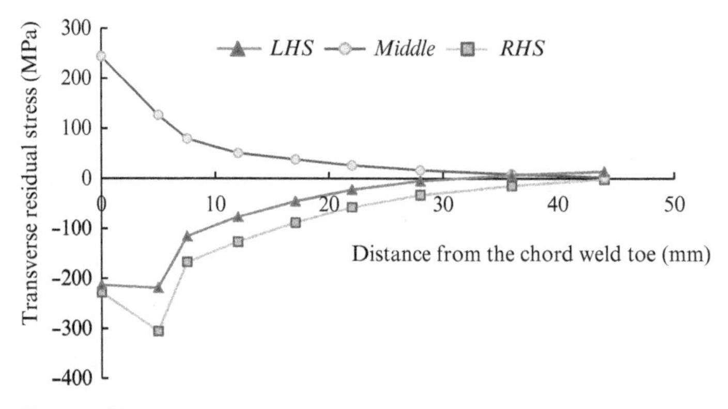

Figure 4.23 Transverse residual stress variation at different locations (θ = 135°, t_1 = 12 mm, ambient temperature, welding direction: case 1).

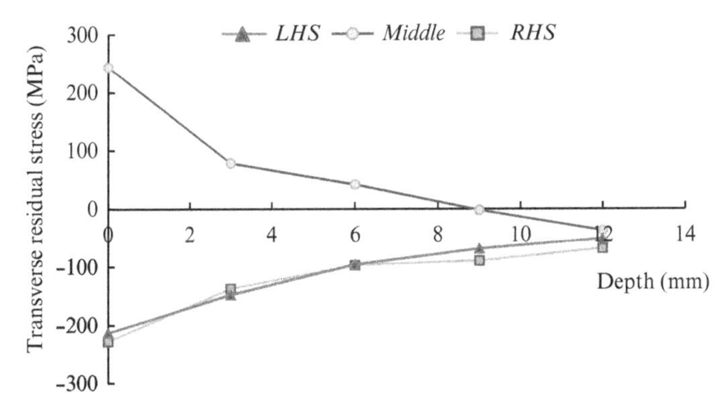

Figure 4.24 Transverse residual stress variation at different depths ($\theta = 135°$, $t_1 = 12\,\text{mm}$, ambient temperature, welding direction: case 1).

Figure 4.23 shows the transverse residual stress variations with the distance from the chord weld toe for the joint welded at ambient temperature (welding direction: case 1). Three cross sections (two width ends, LHS and RHS, and the midpoint along the width direction) are selected to show the relationship between the magnitude of residual stress and the distance from the chord weld toe. The magnitude of residual stress within 10 mm is much higher than the other parts. With an increase of the distance, the magnitude of residual stress reduces quickly. When the distance is beyond 40 mm, it fluctuates near zero. The tensile residual stress is located at the midpoint of the chord width, while compressive residual stresses turn out at both ends of the plate width. The residual stress variations along the plate depth are shown at Figure 4.24. It can be observed that the residual stress on the top surface is much higher than the bottom surface.

4.3.5.2 Preheating Joint

Figure 4.25 shows the transverse residual stress for the joint preheated to 100 °C before welding. Similar with the joint welded at ambient temperature, compressive residual stresses turn out at the two ends of the plate width direction, especially in the region under the weld (midpoint in length). It can be observed from Figure 4.25(d) that the tensile residual stress appeared on the top surface of the chord plate. At this cross section, the maximum compressive residual stress, which is located on the top and bottom surfaces of the chord plate, was generated in the middle along plate width. A layered distribution of residual stress can be found at this cross section in depth. The maximum tensile residual stress turns out on the top surface of the chord plate. With an increase in depth at this cross section, the residual stress decreases. Figures 4.25(c) and 4.25(e) show the transverse residual stress at the ends of plate width. Similar with Figures 4.22(c) and 4.22(e), compressive residual stress turns out on the top and bottom of the chord plate.

Figure 4.26 shows the transverse residual stress variations with the distance from the chord weld toe for the preheated specimen (welding direction: case *1*). Three cross sections (two ends and midpoint along plate width direction) are selected to show the relationship between the magnitude of residual stress and the distance from the chord weld toe. Similar with joints welded at ambient temperature, an inverse relationship between the residual stress magnitude and distance can be

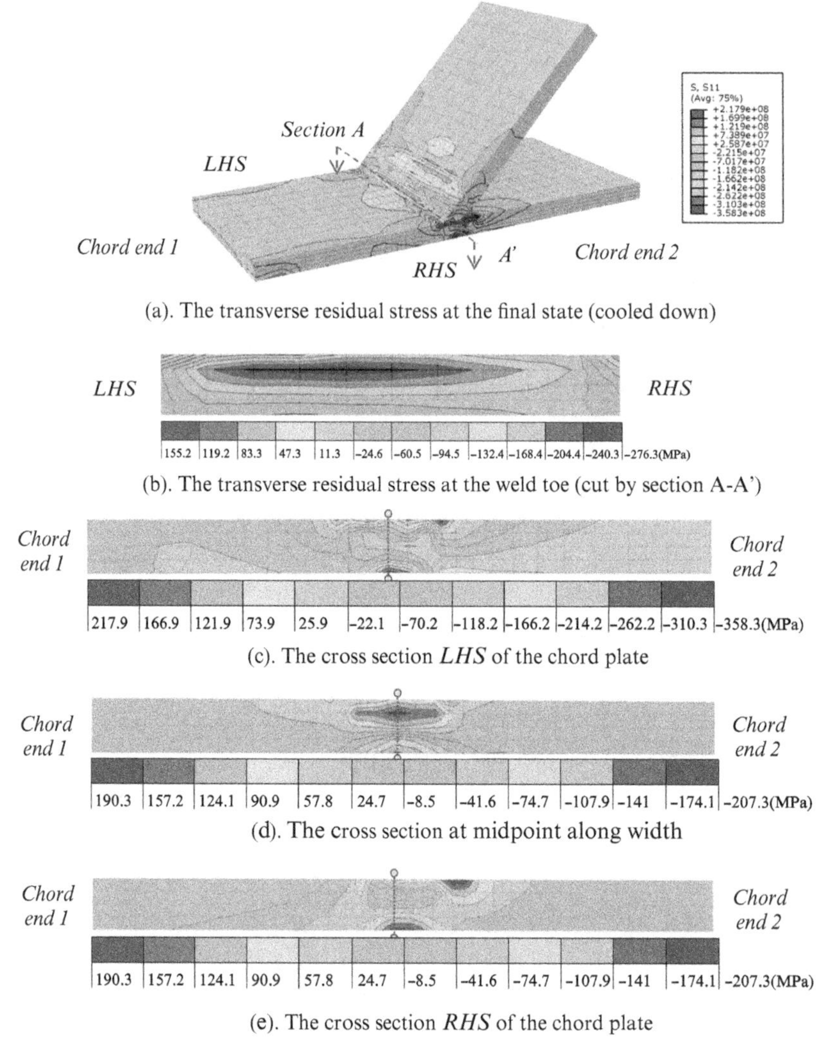

Figure 4.25 The transverse residual stress profile for preheated specimen at the chord plate (*LHS:* left-hand side, *RHS:* right-hand side).

found. The residual stress within 10 mm from the weld toe should be noticed for its high magnitude. When the distance is beyond 40 mm, it fluctuates near zero. The tensile residual stress is located at the middle of chord width, while the compressive residual stresses turn out at the two ends. The residual stress variations along the plate depth are shown at Figure 4.27. It can be observed that the residual stress on the top surface is much higher than the bottom surface.

4.3.5.3 Comparison Between Ambient Temperature and Preheated Joints

Figures 4.28 and 4.29 show the von Mises residual stress at the middle cross section of plate width for the two joints ($\theta = 135°$, $t_1 = 12$ mm, ambient temperature and preheated). Note that the region near the weld toe and root are yielded in this position. But in the internal part of the weld, the magnitude of von Mises residual stress

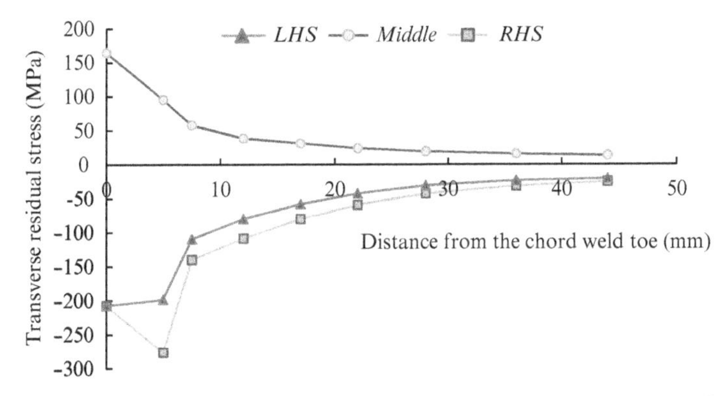

Figure 4.26 Transverse residual stress variation at different locations ($\theta = 135°$, $t_1 =$ 12 mm, preheated, welding direction: *case 1*).

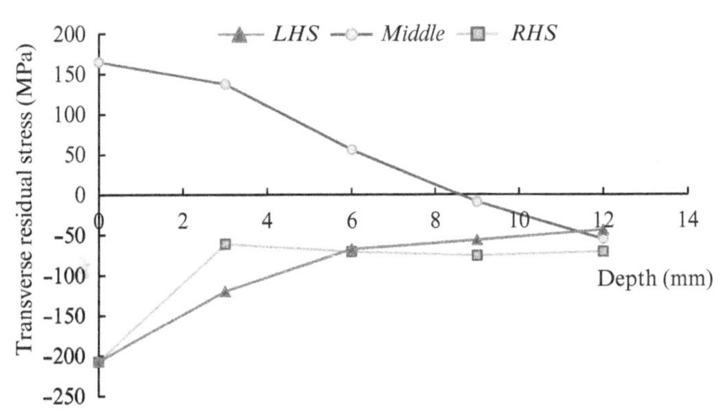

Figure 4.27 Transverse residual stress variation at different depths ($\theta = 135°$, $t_1 = 12$ mm, preheated, welding direction: *case 1*).

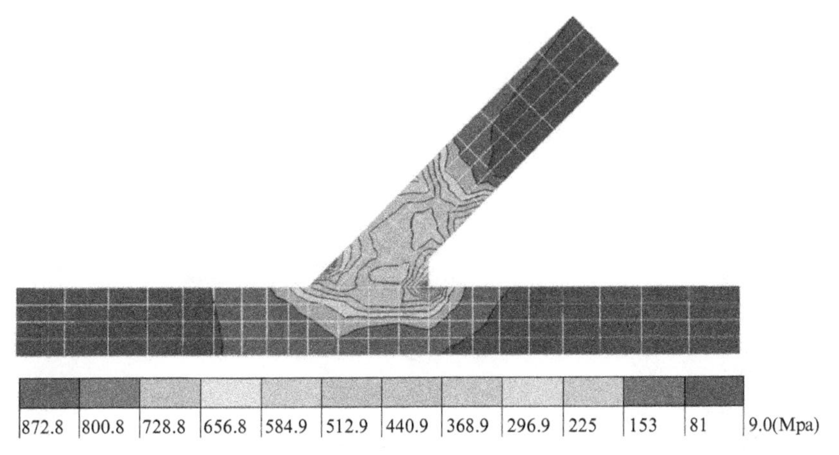

Figure 4.28 The von Mises residual stress in the middle of chord width ($\theta = 135°$, $t_1 =$ 12 mm, ambient temperature, welding direction: *case 1*).

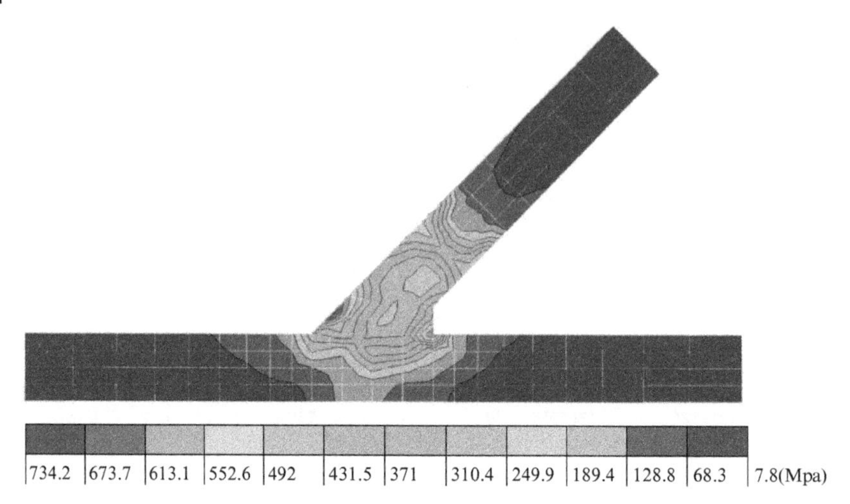

Figure 4.29 The von Mises residual stress in the middle of chord width ($\theta = 135°$, $t_1 = 12$ mm, preheating, welding direction: *case 1*).

is much lower. For the base plate far away from weld, the von Mises residual stress is quite low (less than 50 Mpa). Comparatively, the von Mises residual stress for the preheated joint is shown in Figure 4.29. Compared with the ambient temperature joint, the maximum residual stress is lower for the preheated joint. Note that the maximum von Mises residual stress for the preheated joint is 734.2 MPa, whereas 872.8 MPa can be found in the ambient temperature joint. In addition, the area of high stress gradient near the weld toe and root is reduced as a result of preheating.

To capture the differences between the transverse residual stresses for both joints, Figures 4.30 and 4.31 give the distributions of the transverse residual stress for the middle sections near the weld when they are fully cooled down. Similar stress distribution patterns can be found for the two joints, whereas noticeable residual stress differences can be observed. The maximum transverse residual stress for the ambient temperature joint and the preheated joint are respectively 293.6 MPa and

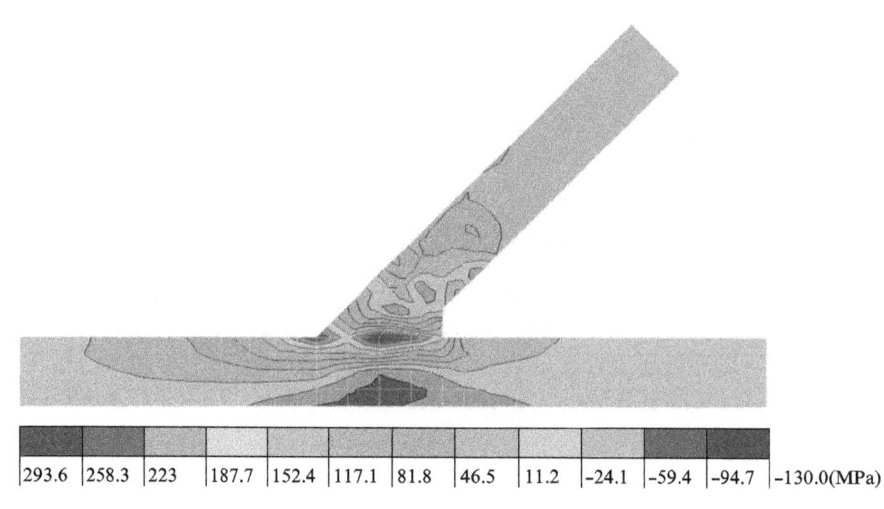

Figure 4.30 The transverse residual stress at the middle of chord width ($\theta = 135°$, $t_1 = 12$ mm, ambient temperature, welding direction: *case 1*).

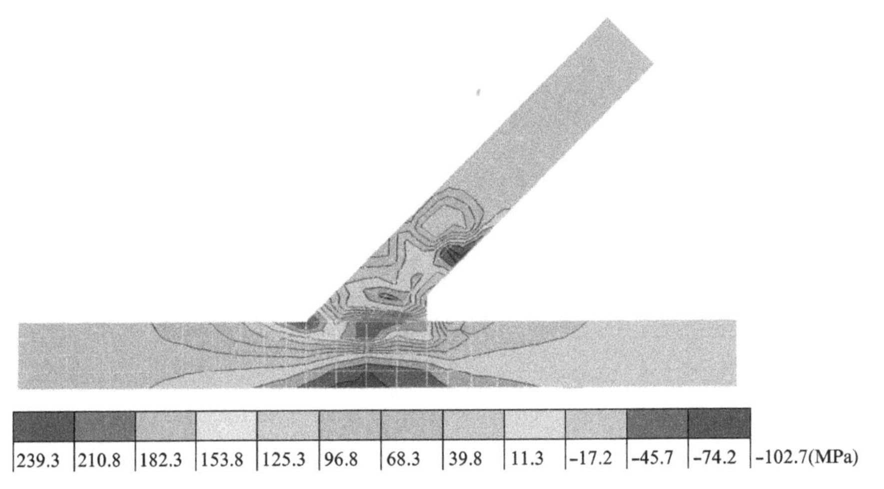

Figure 4.31 The transverse residual stress at the middle of chord width ($\theta = 135°$, t_1=12 mm, preheating, welding direction: *case 1*).

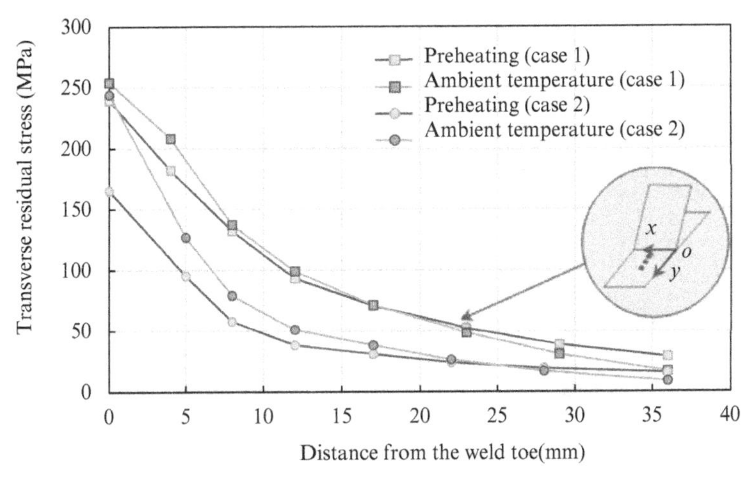

Figure 4.32 Transverse residual stress variation at different locations.

239.3 MPa. The conclusion that preheating can effectively release part of residual stress magnitude for this HSS joint can be drawn. Figure 4.32 shows the comparison between preheated and ambient temperature joints for two welding directions (*case 1* and *case 2*). A reduction of residual stress due to preheating can be found for both cases when compared with ambient temperature. Similar observation can be found for the case ($\theta = 135°$, $t_1 = 12$ mm, ambient temperature, welding direction: case *b*), which is shown from Figure A1.1 to Figure A1.4 of Appendix 1.

4.4 Parametric Study

Following the modelling validation, a small-scale parametric study will be carried out in this section based on 2D models. In this section, 23 models are run and Table 4.3 lists the details of them. The mechanical boundary condition of the joints, the

Table 4.3 Summary of the modeling for parametric study (total: 23 models).

Cases			Residual stress computed at selected points for parametric study						
Parameter considered		Modeling conditions	0 mm	5 mm	10 mm	15 mm	20 mm	25 mm	30 mm
The boundary condition	Fix	Ambient temperature	293.3	80.3	36.5	31.5	27.9	32.6	32.5
	Pin	Weld speed: 2.6 mm/s	341.9	190	120.6	115.7	112.6	110.7	112.3
	Simply support	Weld lump: 4 Weld sequence: a	282.6	76.8	32.1	30.5	28.4	27.6	27.4
Preheating temperature	30 °C	Boundary condition: pin	293.3	80.3	36.5	31.5	27.9	32.6	32.5
	75 °C	Weld speed: 2.6 mm/s	241.2	95.7	45.3	37.1	33.7	26.7	30.7
	100 °C	Weld lump: 4	224	101.4	47.8	38.8	35.3	34	29.2
	150 °C	Weld sequence: a	155.2	96.1	58.7	41.5	36.6	34.4	34.5
	200 °C		112.8	81.9	57.3	47.7	37.5	34.4	34.3
	300 °C		66.2	67.4	41.9	37.5	35.1	36.9	38.0
Welding lumps	2	Ambient temperature	336.7	87.4	34.8	30.1	28.2	31.4	33.6
	4	Boundary condition: pin	293.3	80.3	36.5	31.5	27.9	32.6	32.5
	8	Weld speed: 2.6 mm/s	263.6	77.5	39.4	33.7	31.4	28.7	32.7
	16	Weld sequence: a	241.7	70.8	40.6	33.1	30.7	34.6	35.7

Welding speed	2.0 mm/s	Ambient temperature	360.6	133.7	65	46.6	41	37.4	33.6
	2.2 mm/s	Boundary condition: pin	310.6	104.8	53.4	41.4	36.8	31.8	34.7
	2.4 mm/s	Weld lump: 4	307.2	100	43.7	36.8	31	32.5	34.6
	2.6 mm/s	Weld sequence: a	293.3	80.3	36.5	31.5	27.9	32.6	32.5
	2.8 mm/s		225.8	75.3	30.1	25.4	28.7	30.9	30.2
	3.0 mm/s		174.9	77.9	22.9	23.2	28.5	28.9	28.1
Welding sequence	a	Ambient temperature	293.3	80.3	36.5	31.5	27.9	32.6	32.5
	b	Boundary condition: pin	307.7	95.3	36.2	33.6	30.4	32.7	34.2
	c	Weld speed: 2.6 mm/s	335.9	102.8	42.6	38.2	35.6	32.7	33.5
	d	Weld lump: 4	402.5	128.7	55.6	47.3	45.1	43.2	43.6

preheating temperature, the number of weld lumps, the welding speed, and the welding sequence are evaluated for their effects on the residual stress distribution. In order to keep the number of models within a manageable limit, each parameter is analysed separately by keeping other parameters constant, and only the joint with 135° and 12 mm base plate thickness was studied. Figure 4.33 shows the three mechanical boundary conditions of the joint employed in the parametric study. Figure 4.34 lists the four weld schemes employed and Figure 4.35 shows the four weld sequences in the parametric study. Note that case *b* shown in Figure 4.35 corresponds to the actual welding sequence employed during the fabrication of the joints.

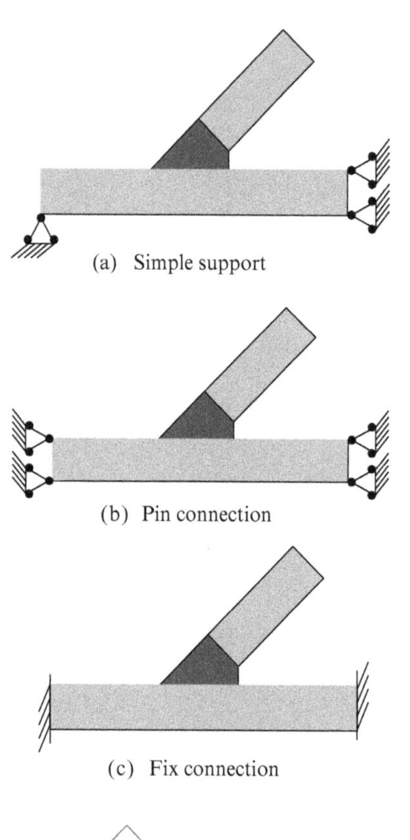

Figure 4.33 Three boundary conditions included in the modeling.

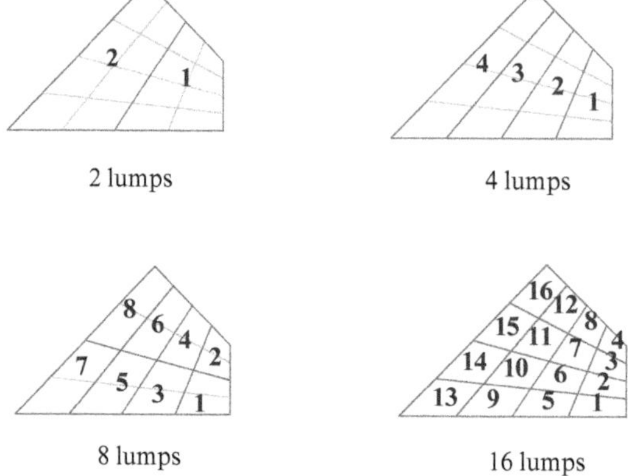

Figure 4.34 Different lumping scheme in the parametric study.

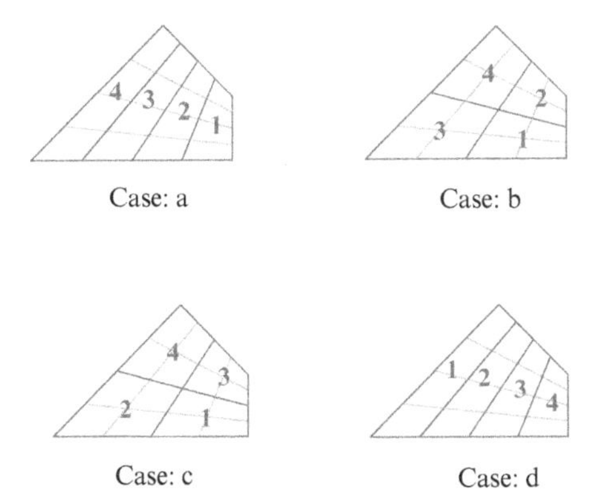

Figure 4.35 Cases for different weld sequence in the parametric study.

4.4.1 Effect of Boundary Condition

Figure 4.36 shows the relationship between transverse residual stress and the distance from the weld toe when the different boundary conditions are considered. Three support cases at the two ends of the chord plate are included in the comparison: fix connections, pin connections, and simply support. There exists a 42.6 MPa stress difference at the weld toe between the fix and the pin connections. However, the stress between the simply support is only smaller than the pin connection by 10.7 MPa at the weld toe. By comparing the modeling results with test results, it is found that simply support can eliminate parts of the residual stress magnitude. However, the angular distortion and deformation in the joint would be more serious for simply support in both ends of the chord plate.

4.4.2 Effect of Preheating Temperature

Figure 4.37 shows the relationship between the transverse residual stress and the distance from the weld toe when the different preheating temperatures were applied.

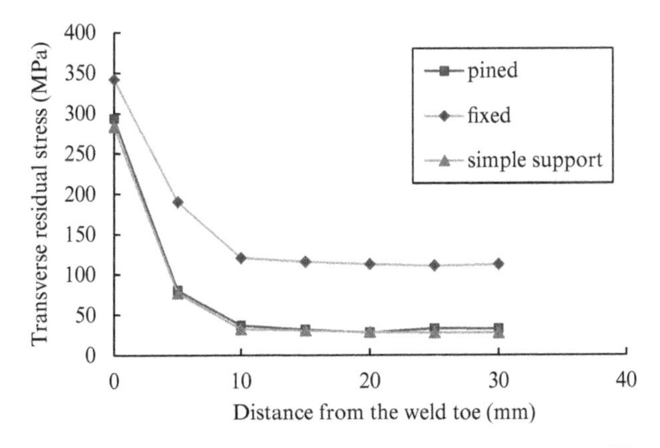

Figure 4.36 Comparison of transverse residual stress under different boundary conditions.

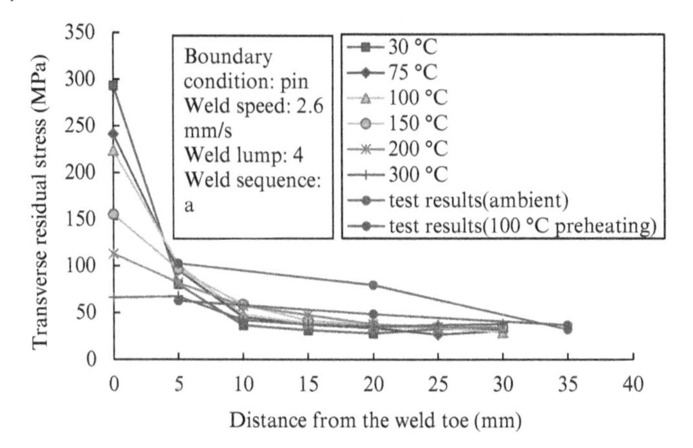

Figure 4.37 Comparison of transverse residual stress under different preheating temperatures.

The preheated areas are within 30 mm of the welding connection (Figure 4.3). During modeling, the preheating effect was added as a pre-defined constant temperature field in the model before the first weld lump was added in. From Figure 4.37, it can be seen that the residual stress value at the weld toe is sensitive to the preheating temperature. When the preheating temperature was increased from 75 °C to 300 °C, the transverse residual stress was dropped from 241.2 MPa to 66.2 MPa. Note that when the joint is welded in ambient temperature, the transverse residual stress at the weld toe is 293.3 MPa. From Figure 4.37, it can be again concluded that preheating can effectively relieve residual stress near the weld toe. Furthermore, Figure 4.37 also reconfirms that such reduction effect will not be significant for regions located beyond 15 mm from the weld toe.

Figure 4.38 shows the variation of the average cooling rate when different preheating temperatures were applied. When comparing Figure 4.37 with 4.38, a similar trend can be observed that at the weld toe, the average cooling rate is the largest

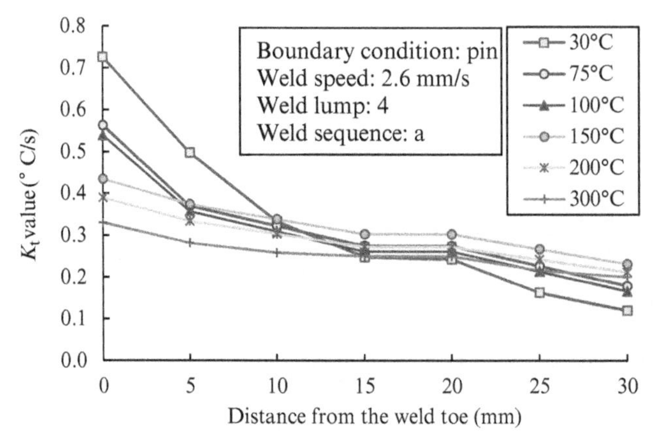

Figure 4.38 The cooling rate at selected points for preheating effect.

when the joint was fabricated at the ambient temperature and the average cooling rate was reduced as the preheated temperature was increased. In addition, the average cool rate dropped quickly with the increase of the distance from the weld toe when the perheating temperature is lower than 150 °C.

Figure 4.38 shows the average cooling rate K_t at the selected points when different preheating temperatures are applied. At the weld toe, the average cooling rate value is largest for the ambient temperature case. At this point, when the preheating temperature was increased, the average cooling rate reduced. Note that when the distance is less than 10 mm, the K_t value reduces significantly with the increase of the distance from the weld. This phenomenon is consistent with the residual stress variations that are shown in Figure 4.37. It can be concluded that preheating can effectively reduce the average cooling rate near the weld part and thus relieves part of the residual stress in this area.

4.4.3 Effect of Using Different Lumps

In order to find out the influence of the number of weld lumps used during modeling on the predicted value of residual stress, the residual stress distributions obtained for four different lumping schemes with 2 to 16 lumps were computed and the results are shown in Figure 4.39. From Figure 4.39, it can be concluded that the predicted residual stress at the weld toe was reduced as the number of weld lumps employed was increased. For the joint studied, the residual stress at the weld toe was decreased from 336.7 MPa to 263.6 MPa as the number of weld lumps was increased from 2 to 16. In particalar, when the numbers of lumps was increased from 2 to 8, the magnitude of residual stress at the weld toes was reduced significantly (74 MPa) while only a small drop (8.2 MPa) occurred when the numbers of lumps was increased from 8 to 16. Based on the above results, it is suggested that in practice, 4 weld lumps should be applied to plate-to-plate HSS joint modeling in order to obtain a residual stress prediction with reasonable computational resources.

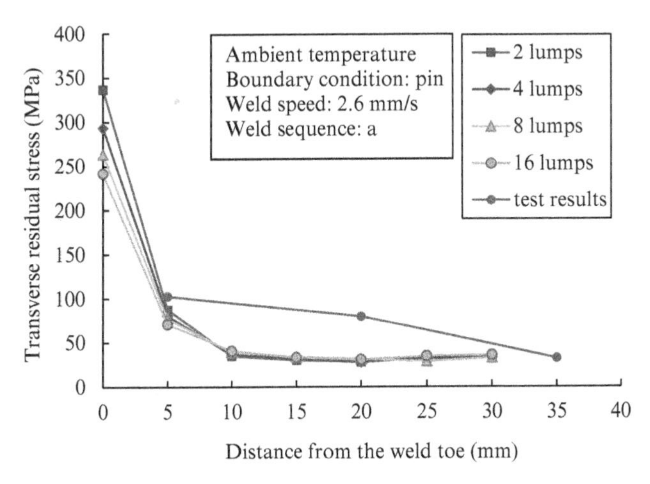

Figure 4.39 Comparison of transverse residual stress under different weld lumping schemes.

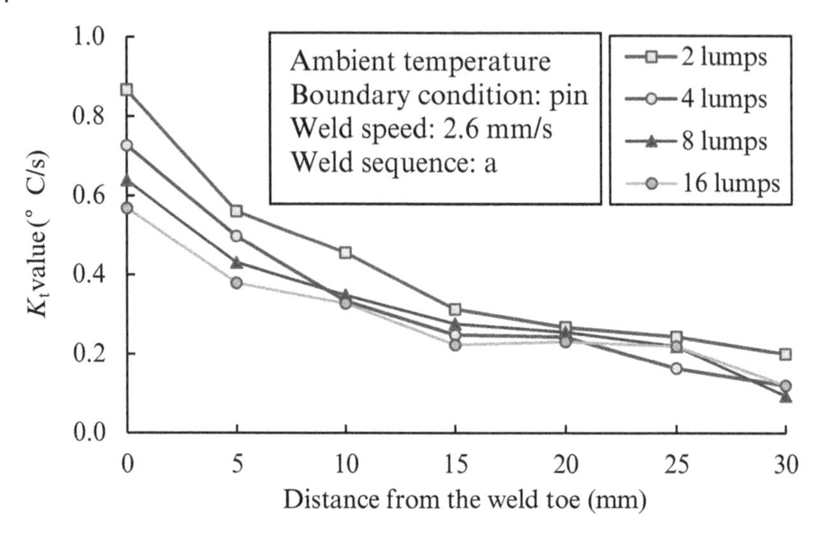

Figure 4.40 The average cooling rate at selected points for different lumping schemes.

Figure 4.40 shows the varation of the average cooling rate when different lumping schemes were used. Similar to the Figure 4.39, again it can be seen that the avarage cooling rate dropped as more weld lumps were employed during the modeling.

4.4.4 Effect of Welding Speed

It should be mentioned that during the parametric study of welding speed, it was assumed that the same welding equipment with the same settings were used when the welding speed was varied. Hence, both the current and voltage applied were kept constant. From Eq. 4.2 (Section 2.4), this implies that as the welding speed is increased, the value of heat flux will be decreased. Together with the obvious fact that a higher welding speed implies a shorter heating time of the section under consideration, it can be concluded that when a higher welding speed is applied, less heat input per unit length will be passed into the joint and it is likely to decrease the magnitude of the residual stress. Figure 4.41 shows the influence of welding speed on the transverse residual stress distribution. As expected, the residual stress decreased as the welding speed increased. In addition, Figure 4.41 shows the residual stress at the weld toe was sensitive to welding speed when it was slower than 2.6 mm/s. When the weld speed was 2.0 mm/s, 2.2 mm/s, and 2.4 mm/s, the corresponding residual stresses at the weld toe were 351.7 MPa, 341.3 MPa, and 318.7 MPa. When the weld speed was increased to 2.8 mm/s and 3.0 mm/s, the stress decreased to 242.6 MPa and 187.6 MPa, respectively. Figure 4.42 shows the variation of the average cooling rate when different welding speeds were applied in the modeling. Again, similar to Figure 4.41, the average cooling rate near the weld toe decreased as the welding speed increased.

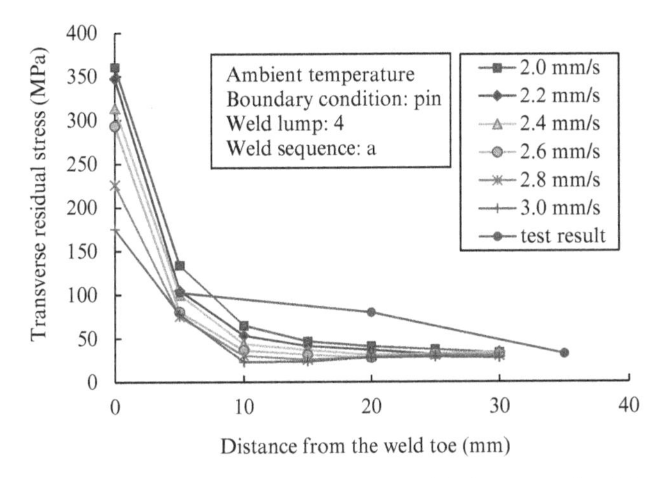

Figure 4.41 Comparison of transverse residual stress under different welding speeds.

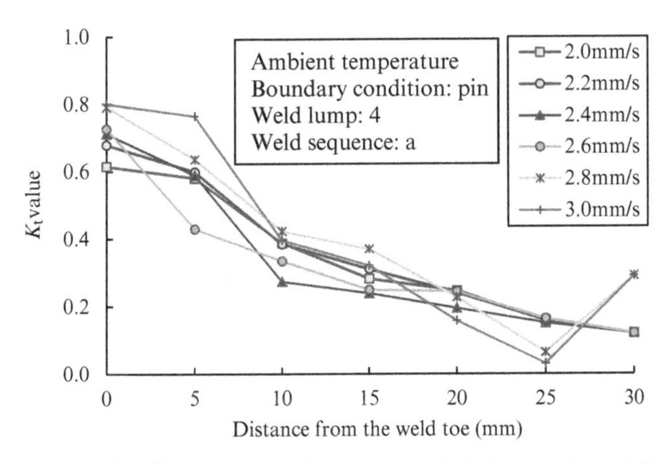

Figure 4.42 The average cooling rate at selected points for welding speed effect.

4.4.5 Effect of Welding Sequence

Four different welding sequences, as shown in Figure 4.35, were studied to evaluate their influence on residual stress in this section. Case *a* is obtained by following the actual weld sequence that is used in joint fabrication. Figure 4.43 shows the relationship between the transverse residual stress and the distance from the weld toe when different welding sequences are applied. When the sequence of the weld filler addition is changed, the magnitude of residual stress, especially at the weld toe, is significantly changed. The stress difference at the weld toe for Case *a* and Case *d* is 104 MPa, though there is not much difference when the distance from the weld toe is larger than 15 mm. It can be concluded that the sequence case *a* can relieve the stress at the weld toe. This phenomenon may be due to the fact that, in the

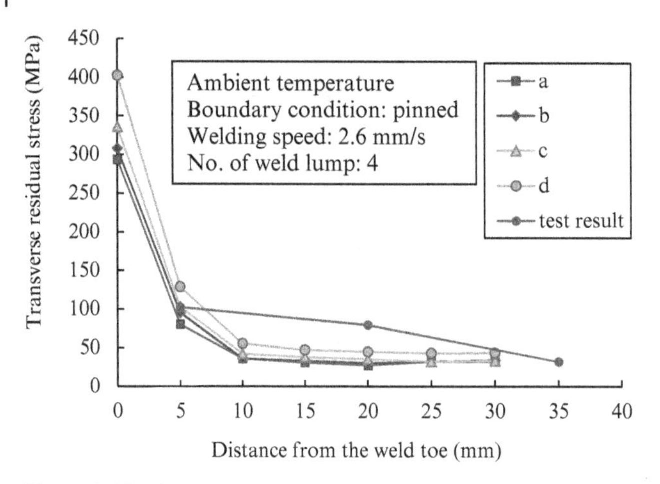

Figure 4.43 Comparison of transverse residual stress under different welding sequences.

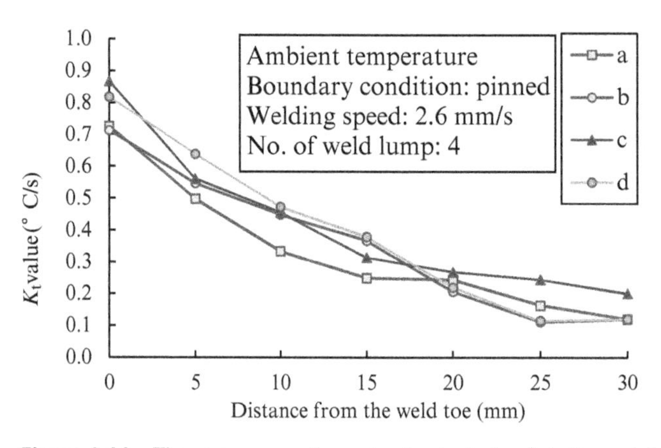

Figure 4.44 The average cooling rate at selected points for welding sequence effect.

multi-pass welding, the first welding pass has a preheating effect for the following welding process. When the sequence case *d* is applied, the weld toe is heated and temperature rises quickly. When the weld fillers are totally added, the cooling rate at the weld toe becomes lower so that part of residual stress can be relieved, as shown in Figure 4.44.

4.5 Conclusions

This chapter presented both 2D and 3D procedures of sequentially coupled thermal-stress analysis for a set of HSS plate-to-plate welded joints. Element birth and death is an effective way to simulate the addition of filler in the procedure of welding. Sequential coupled thermo-mechanical is an acceptable way in the simulation of residual stress.

A small-scale parametric study was carried out to evaluate the influence of the mechanical boundary condition, the preheating temperature, the number of welding lumps, the welding speed, and the welding sequence on the residual stress field. The mechanical boundary condition has influence on the residual stress at the weld toe. When both ends of the chord plate are fixed, the transverse residual stress at the weld toe is higher than that when pin connection is applied on both ends. When the distance from the weld toe goes further from 15 mm to 50 mm, this influence decreases gradually.

In particular, the preheating treatment can reduce the magnitude of the principle residual stress at the weld toe. This effect is more obvious when the preheating temperature rises to 300 °C. The number of weld lumps is a significant parameter that can affect the residual stress distribution near the weld toe. When more weld lumps are applied for a multi-passes welding, a more accurate residual stress at the weld toe can be obtained. Weld speed has significant influence on the magnitude of residual stress at the weld toe. Lower speeds can effectively relieve the residual stress at the weld toe, especially when the weld speed is lower than 2.4 mm/s. Also, when weld filler is added in different sequences, slight changes exist for the magnitude of residual stress. The residual stress at the weld toe is lower when the joint angle increases.

5

Experimental Investigation of Residual Stress for Welded Box High-Strength Steel T-Joints

5.1 Introduction

Compared with the HSS plate-to-plate joints (studied in Chapters 3 and 4) in which the welding is carried out along the chord width direction in the middle of the chord plate, the fabrication procedure of HSS box joints is much more complicated. Since the HSS RQT701 is supplied in the form of a steel plate, when this kind of HSS is used in the tubular structure the box chord and brace hollow sections need to be fabricated first. Therefore, the welding residual stress exists in the chord and brace hollow sections before the welding at the intersection of the brace and the chord boxes is carried out. When welding is carried out at the joint intersection, the final residual stress field in the joint includes two parts, namely the residual stress due to welding in the box forming procedure and the residual stress due to welding the joint intersection. These two residual stresses will superimpose and cause the final residual stress distribution to complicate. This welding residual stress has significant influence on the mechanical behavior of HSS tubular structure. In particular, when a HSS box joint is subjected to cyclic loads, the welding residual stress field can change the stress range, therefore it is necessary to investigate the residual stress formed in the fabrication.

Some investigations on residual stress distribution of hollow sections and tubular joints can be found in the existing literature. Wikandert et al. carried out experimental and numerical investigation for the welding residual stress of a welded D-shaped cross section. Payne et al. (Payne and Porter-Goff 1986) organized experimental residual stress distributions in welded tubular T-nodes. Jang et al. (Jang et al. 2007) analyzed the residual stress distribution near the weld toe of a circular T-joint. One characteristic in common of their works is that they only studied the residual stress caused by welding at joints intersection. In other words, these joints are made of rolled hollow sections. Furthermore, the joints studied in the literature are mostly made of mild steel. The data related to the study of residual stress for welded box HSS joints is scare. Therefore, to understand the welding residual stress in HSS joints, experimental work to determine residual stress distributions near the chord weld toe of two welded HSS box T-joints is described in this chapter.

Welded High Strength Steel Structures: Welding Effects and Fatigue Performance, First Edition. Jin Jiang.

In this chapter, two specimens with the same geometrical size are fabricated to investigate the welding residual stress distributions of HSS box joints. One specimen is welded at ambient temperature while the other one is preheated to 100 °C before welding. Two stages are included in the procedure of specimen fabrication. First, the brace and chord boxes are both formed by flux-cored arc welding (FCAW) by following the American Welding Society (AWS) standard, from four RQT 701 HSS plates whose nominal yield stress equals 690 MPa. Second, the brace and chord boxes are welded together to form the T-joints. The residual stress near the chord weld toe of the joints is measured by the incremental hole-drilling method according to the ASTM standard E837-08.

5.2 Experimental Investigation

To measure the variation of the residual stress near the chord weld toe, the increased hole-drilling method was used by following the ASTM E837-08 standard (ASTM 2008). The residual stress was calculated by releasing the localized stress and thus changing the local strain on the surface of the testing when a small hole was drilled in a residually stressed structure. The fabrication procedure of the specimens, testing routine, and detailed results are presented in the following sections.

5.2.1 Material Properties

The material properties of the base steel and electrode used in the specimens' fabrication are introduced in this section. As mentioned in Chapter 3, to provide accurate input in modeling and understand the mechanical behavior of this HSS RQT701, the coupon tests at normal and elevated temperature were organized by following the standard (BSI 1992). A special strain gauge, ZC-NC-G1265, which was designed by Vishay Micro-Measurement for high-temperature strain measurement, was used during the coupon test to monitor strain variation under tensile loading. Table 5.1 gives its mechanical properties at room and elevated temperatures.

The electrode OK Tubrod 15.09 was used during the FCAW. It is formed with cored wire for welding high-strength steel with minimum yield strength of 690

Table 5.1 Material properties at room and elevated temperatures.

Temperature	20	100	200	300	400	500	600	700	800
Young's modulus (GPa)	201.3	202.7	205.7	192.3	202.7	154.9	131.9	90.4	57.6
Yielding stress (MPa)	770	757	745	734	675	173	94	40	43
Ultimate stress (MPa)	909	871	847	865	727	428	171	99	66

MPa. The diameter of the electrode is 1.2 mm. Table 5.2 lists the detailed welding parameters that are used in the specimen fabrication.

5.2.2 Specimen Fabrication

5.2.2.1 Overview of the Welding Design

Two HSS box T-joints are fabricated with FCAW welding, which was carried out by Yongnam Holdings Limited, Singapore. In the welding process, the AWS D1.1 2008 (AWS 2008) standard was followed to obtain full penetration weld. To evaluate the impact of preheating on residual stress, one specimen was preheated to 100 °C while the other joint was at ambient temperature before welding. Figure 5.1 shows the dimensions of the box T-joints studied in this chapter.

The fabrication of the HSS box joints can be divided into two stages. The first stage was to fabricate the welded chord and brace boxes from the HSS plates. The second

Table 5.2 Welding parameters for box sections and joints.

		Electrode		Current			
Position	**Process**	**Class**	**Diameter**	**Type**	**Amplitude**	**Volts**	**Inter Pass Temperature**
Box section	FCAW	E111T1-K3	1.2 mm	DCEP	300–320	30–32	170–210
T-joint	FCAW	E111T1-K3	1.2 mm	DCEP	300–320	30–32	170–210

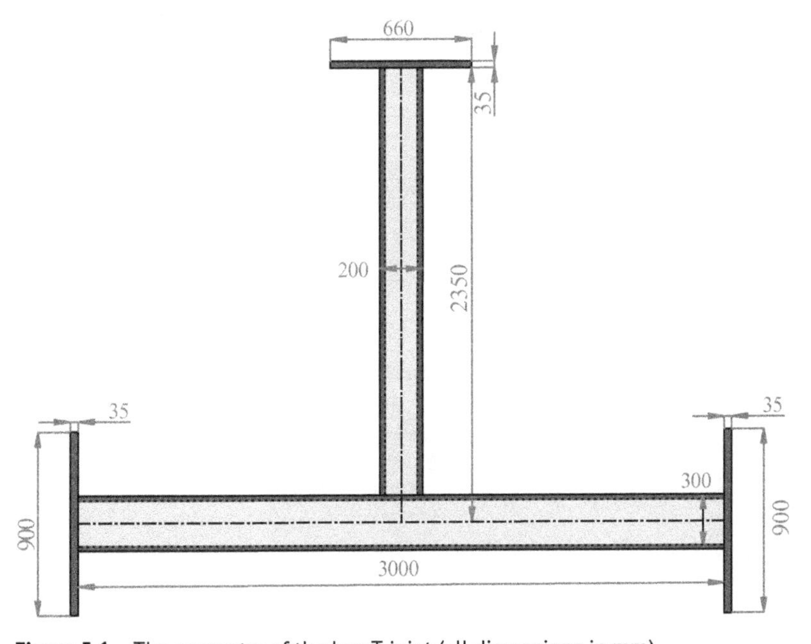

Figure 5.1 The geometry of the box T-joint (all dimensions in mm).

stage was to assemble the chord and brace boxes to the T-joint and carry out the welding in the joint intersection. In the first stage, the RQT701 steel plate was cut into several strips (Figure 5.2(a)). Two sizes of these steel strips (3000 mm×300 mm×12 mm, 2350 mm×200 mm×12 mm) were used respectively for the chord and brace boxes. After that, the brace and chord boxes were assembled as shown in Figures 5.2(b) and 5.2(c). To sustain the molten weld, four backing plates (3000 mm×50 mm×10 mm) with yielding stress equal to 355 MPa were prepared and they were connected to the chord boxes by spot welding (Figure 5.2(c)). A similar preparation procedure was also employed for the brace boxes, except the backing plates size was 2350 mm×50 mm×10 mm. The role of these backing plates was to sustain the weld filler when the welding was carried out to form the box sections. In order to prevent the large deformation of the box section due to the welding, two retaining plates were welded to the box by spot welding, as shown in Figure 5.2(c). When these preparations were done, the molten weld

(a) (b) (c) (d) (e) (f) (g) (h)

Figure 5.2 Fabrication procedure of HSS box T-joint.

was added on the groove to form the brace and chord until full penetration weld was obtained. In the second stage, the T-joint was formed by welding the brace and chord together. Finally, connecting plates used to install the joint into the test rig for fatigue testing were prepared and painting was carried out to prevent corrosion (Figures 5.2(g) and 5.2(h)).

Figure 5.3 shows the layout and welding sequence of the cross section of the brace and chord box sections. Note that welding passes 1 to 4 were sequentially added at four corners of the box section. After that, welding was carried out at each corner one by one until full penetration weld profile was obtained. The box section was fabricated in this manner so that the box sections could be heated evenly to reduce the structural deformation caused by the welding. Figure 5.4 shows the weld profile for the final box joints.

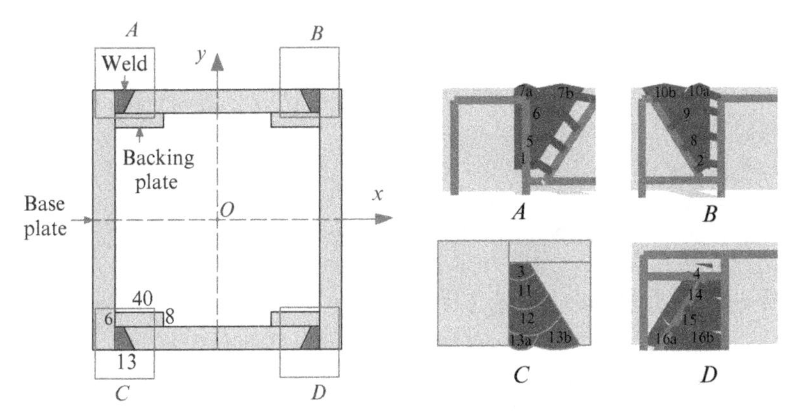

Figure 5.3 The cross section and welding sequence of the box section (all dimensions in mm).

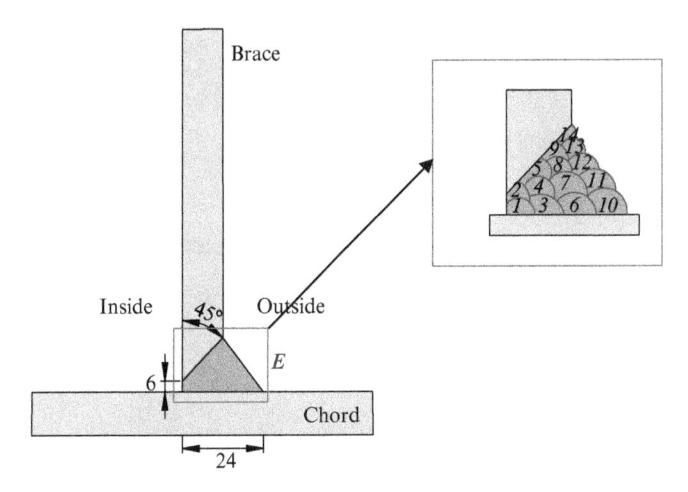

Figure 5.4 The welding sequence of cross section at the intersection (all dimensions in mm).

5.2.2.2 Fabrication of Box Sections

The welding operations were different for the joint welded at ambient temperature and the preheated joint. For the joint welded at ambient temperature, the weld filler was gradually added at the reserved groove without pause until a complete weld pass was finished in the box length direction. However, the procedure for the preheated joint is more complicated. For the preheated joint, one complete weld pass was composed of 6 steps in the welding procedure for the chord box hollow sections. Figure 5.5(a) shows the operation sequence during the welding process. The chord box section was divided into 3 parts with same length (1m). Preheating was firstly carried out at Section 1 (Step 1 in Figure 5.5(a)). The preheating area was roughly 1000 mm×80 mm. When the target temperature (100°) was achieved, the weld filler was added into the groove (Step 2 in Figure 5.5(a)). After that, preheating was carried out at Section 2 and weld filler was added by following in this section. A similar procedure was repeated until a complete weld pass was obtained in the box length direction. For the brace box, a similar preheating procedure was applied, except that it was divided into 2 sections with same length (1m).

5.2.2.3 Fabrication of Joint Intersection

Figure 5.6 shows the welding direction at the chord-brace intersection part for both joints. One complete weld pass is composed of four welding steps. The first step starts from Point 1 and ends at Point 7. The second step is from Point 7 to Point 13, and the third step is from Point 13 to Point 19. Finally, the weld filler starts from Point 19 and returns back to Point 1. Note that the welding starting location is chosen at Point 1. Two two things were taken into account when making this design. One consideration is to reduce the welding residual stress at the corners (the distance from Point 1 and Point 2 is 36 mm). When the molten weld is gradually added from Point 1 to Point 2, the Corner *a* would be heated to an elevated temperature

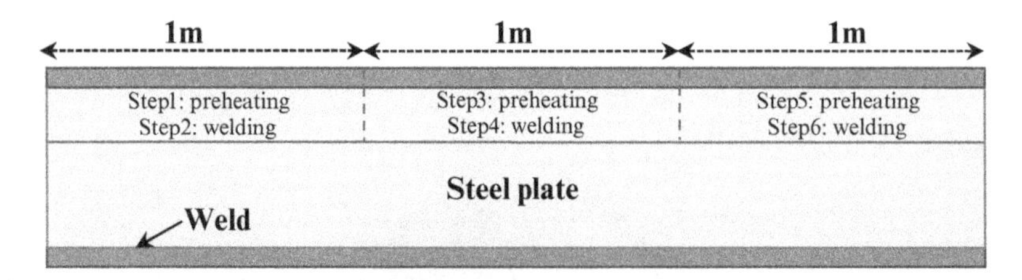

(a) Operation sequence in one weld pass in box length direction

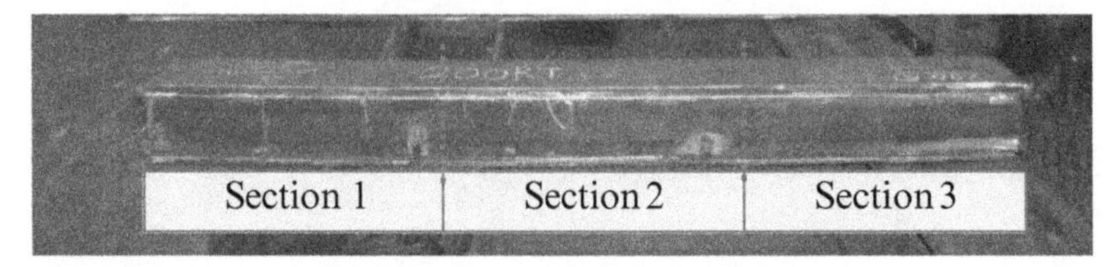

(b) Photo of fabricated box after welding

Figure 5.5 Operation in the welding process for preheated box hollow section.

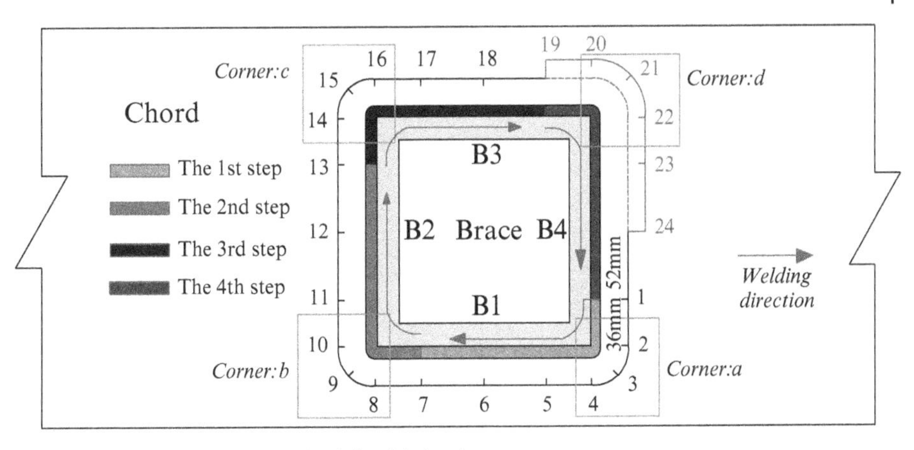

Figure 5.6 Welding direction for joint fabrication.

and it means that there is a preheating effect of added weld filler on the corner. The second reason is to make it easier for the welder to operate the welding process. By doing it this way, the welding would not be paused at the corners of the joints and the welder can adjust his position to achieve a continuous weld pass without many breaks. The same welding process is used for all 14 weld passes (Figure 5.4) until the full penetration weld profile is achieved.

For the specimen designed with the preheating treatment, the area close to where the weld filler was added was heated up to 100 °C before the welding. It was required that the area within 50 mm from the chord weld toe attained a preheat temperature of 100 °C. Temperature chalk was employed to ensure the preheating temperature was reached. After the box formation and joint welding, ultrasonic testing was carried out to ensure the welding quality.

5.2.3 Residual Stress Test Setup and Procedure

The amended RS-200 milling guide was again used to measure the welding residual stress in the HSS T-joints by the hole-drilling method through positioning and drilling of a hole in the center of a special strain gauge rosette. Considering the fact that the original setup (RS-200) made it hard to drill close to the chord weld toe due to blockage of the vertical brace, the milling guide was improved by moving the drilling journal from base plate center to the edge. With this modification, the closest measuring locations can be as close as 5 mm from the chord weld toe. In the drilling process, the strain readings are recorded for every 0.05 mm depth until a 1 mm hole with diameter of 1.6 mm is obtained. The strain readings at 20 different hole depths were recorded to find the distribution of welding residual stress along plate depth (Figure 5.7).

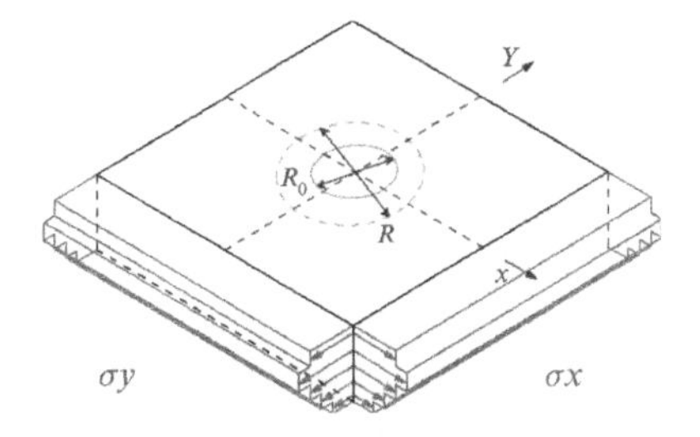

Figure 5.7 Hole geometry and residual stresses.

Before the drilling, the strain gauges, FRAS-2, were attached on the surface of the chord box. Before drilling, accurate alignment is important since even small eccentricity and misalignment between the center of the drilled hole and the target strain gauge FRAS-2 can introduce significant error in the residual stress measurement. Hence, an optical microscope was used to ensure the accuracy of the alignment (less than 0.025 mm). During the drilling, an air turbine was used to drive the tungsten cutter to rotate at very high speed (up to 400 000 *rpm*) to minimize the machine-induced residual stresses at the hole boundary. Figures 5.8 and 5.9 show the amended drilling setup and its close view.

5.2.4 Strain Gauge Schemes for Residual Stress Measurement

For the specimen with preheating, 24 monitoring points (Point 1 to 24 in Figure 5.10) around the brace-to-chord intersection were selected to capture the residual stress distribution near the chord weld toe. Points 1 to 18 are positioned at 10 mm from the weld toe while Points 19 to 24 are positioned at 15 mm from the chord weld toe. A

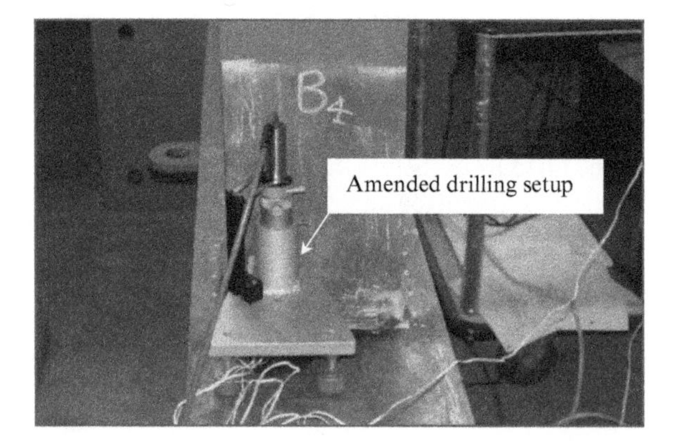

Figure 5.8 Revised drilling setup.

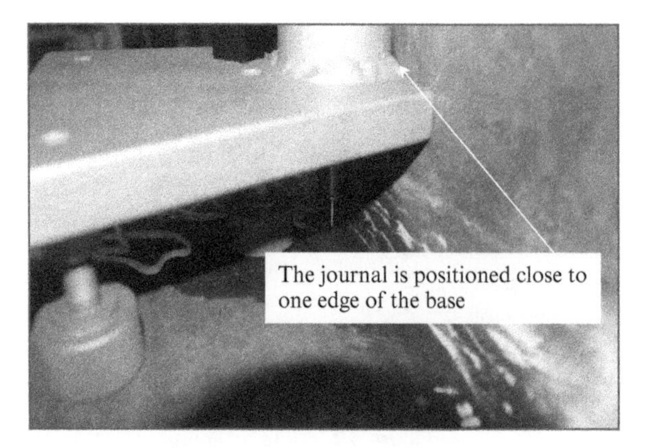

Figure 5.9 Close view of the drilling setup.

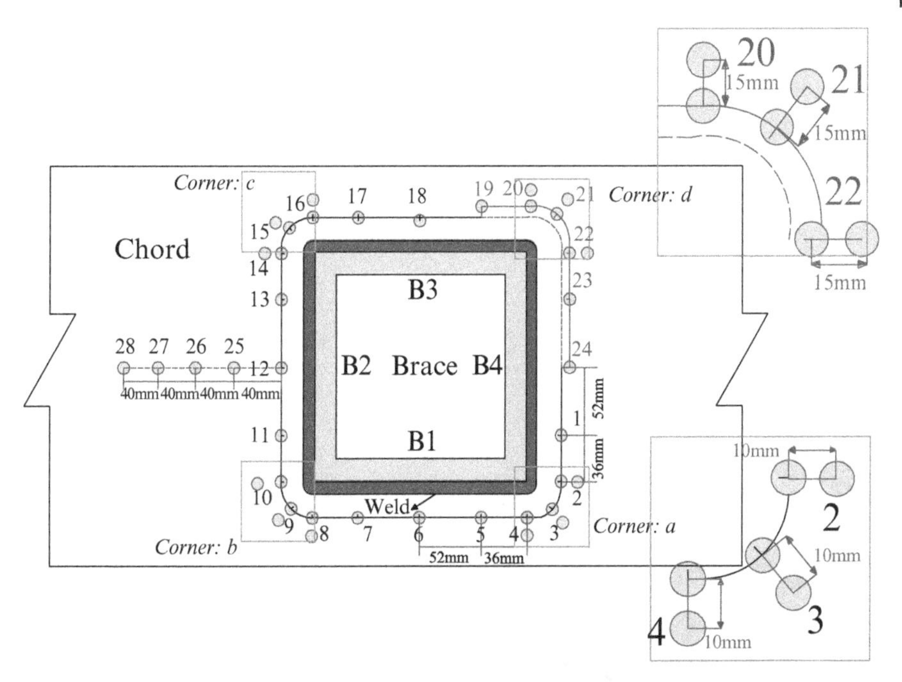

Figure 5.10 The strain gauges for the specimen with preheating.

larger distance was chosen for Point 19 to Point 24 so that there was enough spacing to put ACPD probes in this area for the planned fatigue testing in the future. By this way, the drilling position in this area was further away from the weld toe to minimize the impact of the drilled hole on the fatigue performance. Beside these 24 monitoring points around the intersection, another group of strain gauges was added at 20 mm from the chord weld toe at Corners *a*, *b* and *c* and 25 mm at Corner *d* so that interpolation method could be used to estimate the residual stress at the weld toe.

For the specimen welded at ambient temperature, besides the 24 strain gauges which were positioned the same as the preheating specimen, 4 monitoring points (Points 25–28) were added near Point 12 with spacing of 40 mm to evaluate the residual stress variation with the distance from the weld toe (Figure 5.10). Additionally, another 4 monitoring points (Points 29–32 in Figure 5.11) were added on the top surface of the chord box section. They were aimed to monitor the residual stress in the area close to the weld near the chord section wall at where far away from the weld in the intersection. Note that Points 29 and 30 are on the line with Points 16, 17, and 18. Similarly, the Points 31 and 32 are on the line with Points 3, 4, 5, 6, and 7. Figures 5.12 and 5.13 show the strain gauge schemes at Corners *b* and *d* respectively.

5.2.5 Computation of Residual Stress

The computation procedure of residual stress follows the ASTM 837-08 standard, which is summarized in Figure 5.14. Note that the hole-drilling calibration matrixes in this study are cited from Table 5.5(b) in the standard ASTM 837-08. By adapting

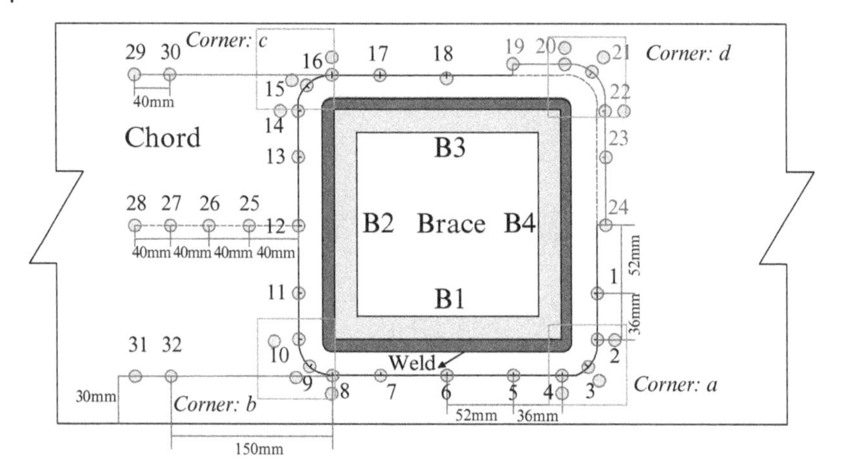

Figure 5.11 The strain gauges for the specimen at ambient temperature.

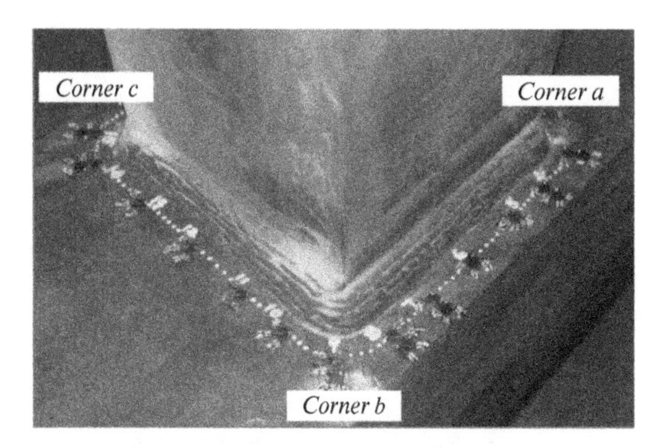

Figure 5.12 Strain gauge scheme around Corner *b*.

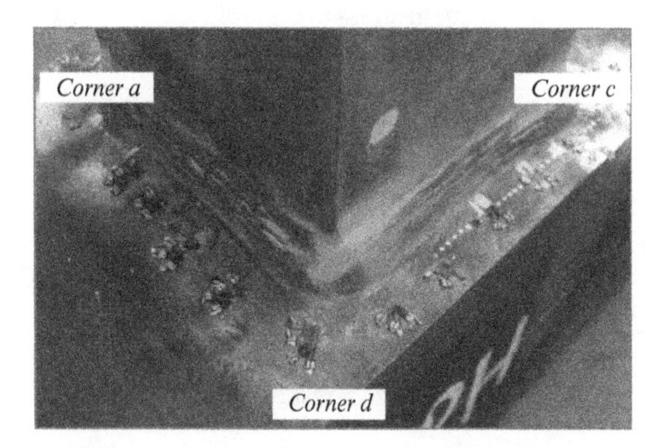

Figure 5.13 Strain gauge scheme around Corner *d*.

this approach, the welding residual stress at different depths can be calculated. In the following sections, when a single stress value is mentioned for a monitoring point, this value is an arithmetic mean value based on 20 different steps measurements within the 1 mm hole.

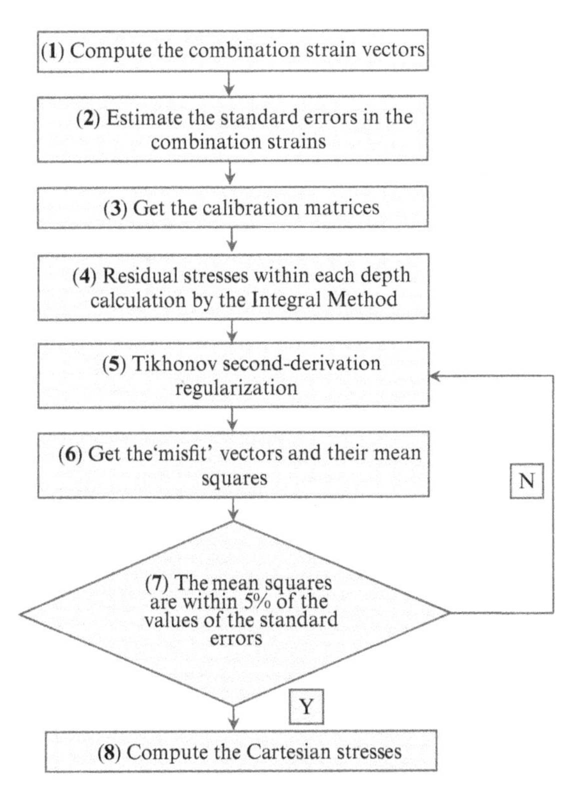

Figure 5.14 Residual stress calculation procedure.

5.3 Testing Results

5.3.1 Preheated Specimen

The transverse residual stress directions at different monitoring points are shown in Figure 5.15. Figure 5.16 gives the distribution of the maximum residual principal stress and its direction at a distance of 10 or 15 mm from the chord weld toe for the preheated specimen. The distance from Points 1 to 18 to the chord weld toe is 10 mm. For Points 19 to 24, this distance is 15 mm. As mentioned, it was planned to put ACPD probes for the future fatigue test. The residual stress shown in the figure is the average value based on 1 mm hole-drilling.

It can be found that the magnitude of residual stress is much higher in the corners than the other points. In particular, at Point 9 and Point 22 the maximum principal residual stresses are respectively 265.5 MPa and 582.4 MPa. Another finding is that the residual stresses along the B2 and B4 sides are higher than the residual stress along the B1 and B3 sides. For principal stress direction, smaller shifting from the transverse direction happens along the B2 and B4 sides when compared with the B1 and B3 sides.

The testing results of the points at 20 or 25 mm from the chord weld toe are shown in Figure 5.17. In general, the principal residual stresses at 20 mm/25 mm are smaller compared with the values at 10 mm/15 mm. However, it should be noticed that at Point 3, Point 4 and Point 7, Point 21, the principal stress at 20 mm/25 mm is higher than at 10/15 mm.

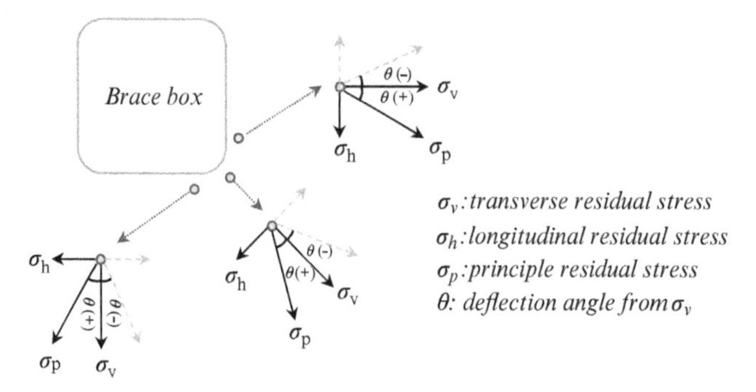

Figure 5.15 Sign conventions for residual stresses.

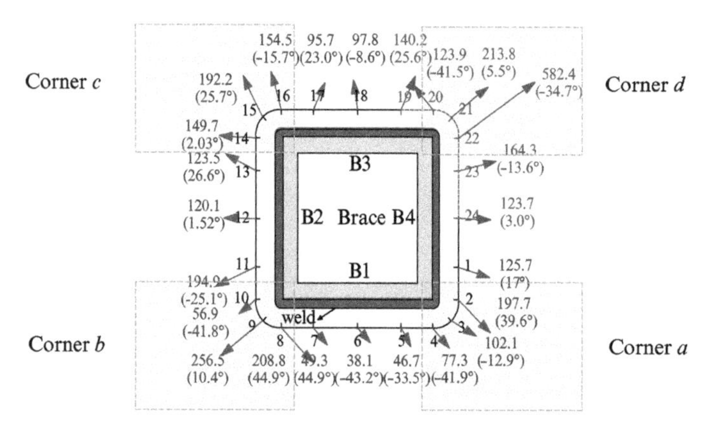

Figure 5.16 The maximum principal stress distribution for the specimen with preheating (position: 10/15 mm, unit: MPa).

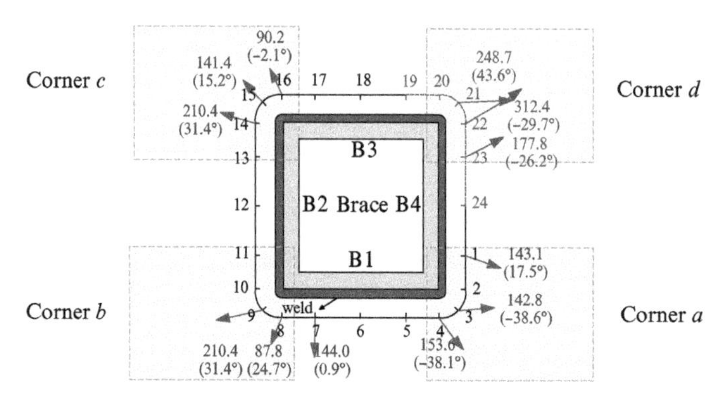

Figure 5.17 The maximum principal stress distribution for the specimen with preheating (position: 20/25mm, unit: MPa)

Considering that the transverse residual stress is related to the fatigue performance of tubular structures, the distribution of the transverse residual stress around the chord weld toe was studied. Figures 5.18 and 5.19 respectively give the distributions of transverse residual stress at 10/15 mm and 20/25 mm. In Figure 5.18, the

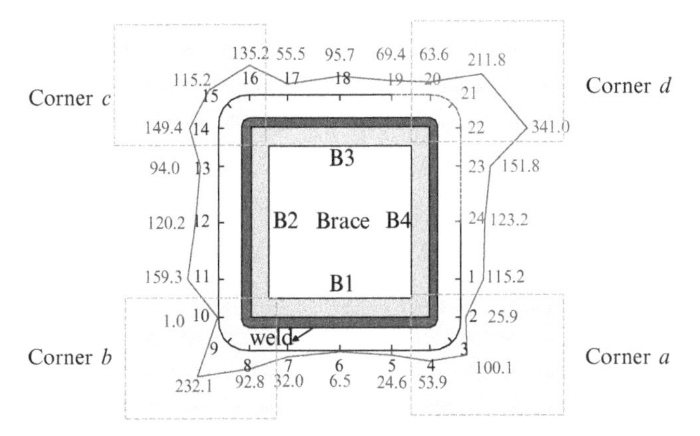

Figure 5.18 The transverse residual stress distribution for the specimen with preheating (position: 10/15mm, unit: MPa).

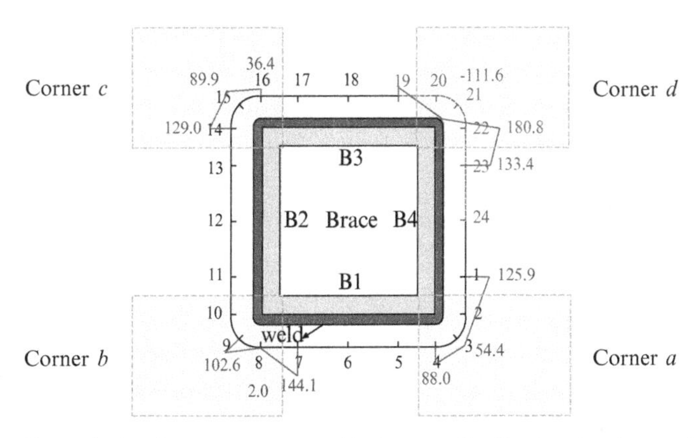

Figure 5.19 The transverse residual stress distribution for the specimen with preheating (position: 20/25mm, unit: MPa).

transverse stresses along B1 and B3 sides are significantly smaller than that along B2 and B4 sides. A notable point in Figure 5.18 is Point 22, where the transverse residual stress is 341 MPa (approximately half of the yielding stress of the material). Another notable point is Point 9, where the transverse residual stress is 232.1 MPa (approximately 1/3 of the yielding stress of the material). In short, high transverse residual stress tends to emerge at the joint corners, especially for Corners *b* and *d*.

For the residual stress at the 20/25 mm, as shown in Figure 5.19, an outstanding position is Point 21, where compressive residual stress is engendered. When compared with the residual stress at 10/15 mm, the magnitude of residual stress at 20/25 mm is much lower generally. However, there are some exceptions, such as Point 4 and Point 7, where a larger residual stress can be found at 20/25 mm than that at 10/15 mm.

The longitudinal residual stresses (*parallel to the welding travelling direction*) for the preheated joint are shown in Figures 5.20 and 5.21. Note that the magnitude of longitudinal residual stress is much lower than the transverse residual stress at a

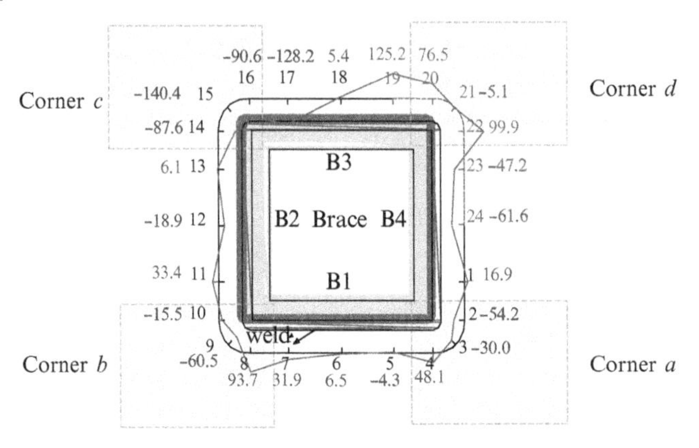

Figure 5.20 The longitudinal residual stress distribution for the specimen with preheating (position: 10/15mm, unit: MPa).

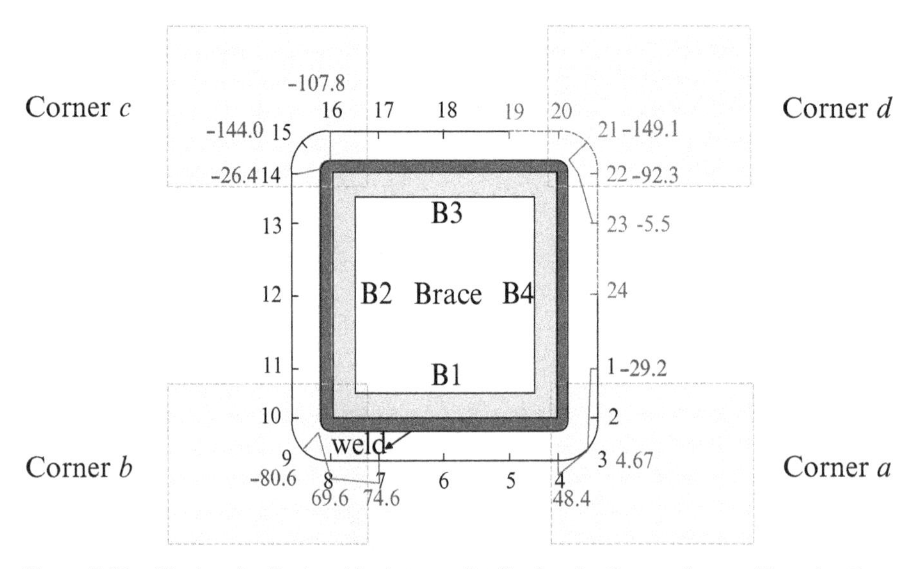

Figure 5.21 The longitudinal residual stress distribution for the specimen with preheating (position: 20/25mm, unit: MPa).

selected point. As shown in Figure 5.20, the longitudinal residual stress around the joint intersection fluctuates near zero. Even the stress values at the four corners are compressive stress. Such similarity also can be found at 20/25 mm points, which is shown in Figure 5.21.

When the residual stress distributions at 10/15 mm and 20/25 mm positions are obtained, the linear interpolation method is used to estimate the transverse and longitudinal residual stress at the chord weld toe. The linear interpolation method is used because the drilling position is hard to move to the chord weld toe and there is not enough space to attach three strain gauges along the B1 and B3 sides. Another consideration is that it can also minimize the influence of drilled holes on the fatigue performance in the future fatigue testing. Figure 5.22 shows the comparison between the transverse and longitudinal residual stress at the chord weld toe. It can be found that the transverse stress is higher than the longitudinal component. This observation is obvious at the corners. Figure 5.23 shows the relationship between the transverse residual stress and the distance from the chord weld toe. With the increase of the distance, the transverse residual stress shows a decreasing tendency. Table 5.3 lists the detail values at measurement points for the preheated joint.

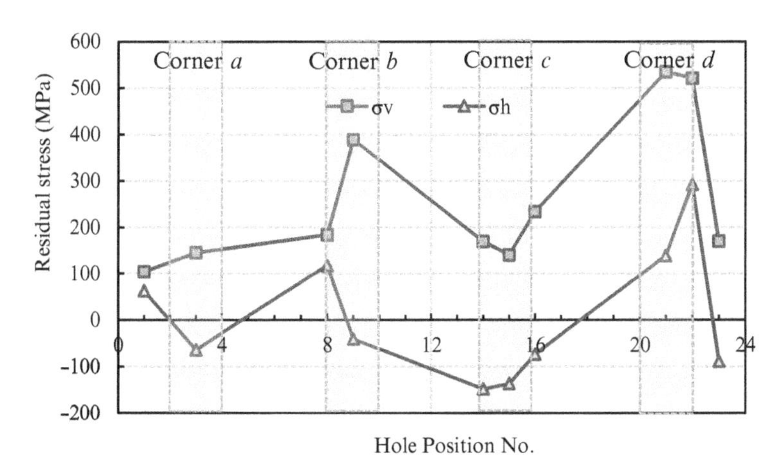

Figure 5.22 The transverse and longitudinal residual stress at the chord weld toe (preheated specimen).

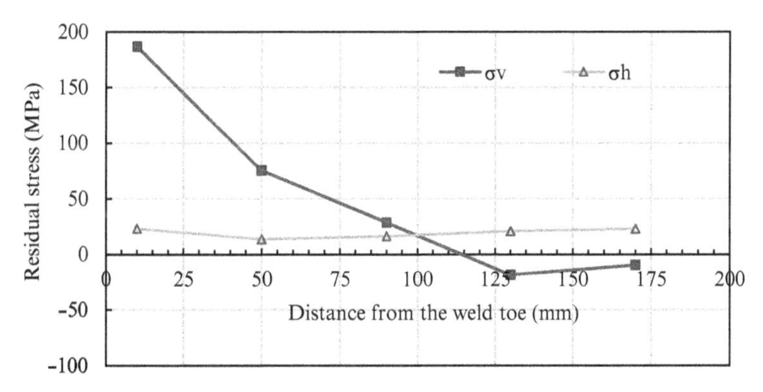

Figure 5.23 The relationship between the transverse and longitudinal residual stress with the distance from the weld toe (preheated specimen).

Table 5.3 Residual stress of the preheated specimen.

Hole No.	σ_v (MPa)	σ_h (MPa)	σ_{max} (MPa)	θ (Figure 5.15)
1	115.2	16.9	125.7	17.0
2	25.9	-54.2	197.7	39.6
3	100.1	-30.0	102.1	-12.9
4	53.9	48.1	77.3	-41.9

(Continued)

Table 5.3 (Continued)

Hole No.	σ_v (MPa)	σ_h (MPa)	σ_{max} (MPa)	θ (Figure 5.15)
5	24.6	−4.3	46.7	−33.5
6	6.5	2.3	38.1	−43.2
7	32.1	31.9	49.3	44.9
8	92.8	93.7	208.8	44.9
9	232.1	−60.5	256.5	10.4
10	1.1	−15.5	56.9	−41.8
11	159.3	33.4	194.9	−25.1
12	120.2	−18.9	120.1	1.5
13	94.0	6.1	123.5	26.6
14	149.4	−87.6	149.7	2.0
15	115.2	−140.4	192.2	25.7
16	135.2	−90.6	154.5	−15.7
17	55.5	−128.2	95.7	23.0
18	95.7	5.4	97.8	−8.6
19	69.4	125.2	140.2	25.6
20	63.6	76.5	123.9	−41.5
21	211.8	−5.1	213.8	5.5
22	341.0	99.9	582.4	−34.7
23	151.8	−47.2	164.3	−13.6
24	123.2	−61.6	123.7	3.0

5.3.2 Ambient Temperature Specimen

Figure 5.24 gives the distribution of the maximum principal stress and its direction for the specimen welded at ambient temperature. Note that the magnitude of residual stress in Corners *b* and *c* is higher than the other points for this specimen. The principal stress at Point 10, Point 11, and Point 14 is respectively 451.6 MPa, 401.2 MPa, and 413.6 MPa. Similar with the preheating specimen, the residual stresses at the points along the B2 and B4 sides are higher than the residual stresses at the points along the B1 and B3 sides. Principal stress direction shifting is higher in the corners than the other positions for this joint. Figure 5.25 shows maximum principal stress and its direction at 20 mm/25 mm at four corners. Note that at Points 4, 8, 16, and 20 (the points located along the B1 and B3 sides, which are close to weld in chord box, see Figure 5.5), this principal stress at 20/25 mm points is higher than the stress at 10/15 mm points.

It is shown in Figures 5.26 and 5.27 for the distributions of the transverse residual stresses at the 10/15 mm and 20/25 mm locations. When compared with the residual stress at the points along the surface of the B2 and B4 sides, the transverse residual stress at the points along the surface of the B1 and B3 sides is lower. The maximum transverse residual stress is 412 MPa at 10 mm, located at Point 14. It also should be

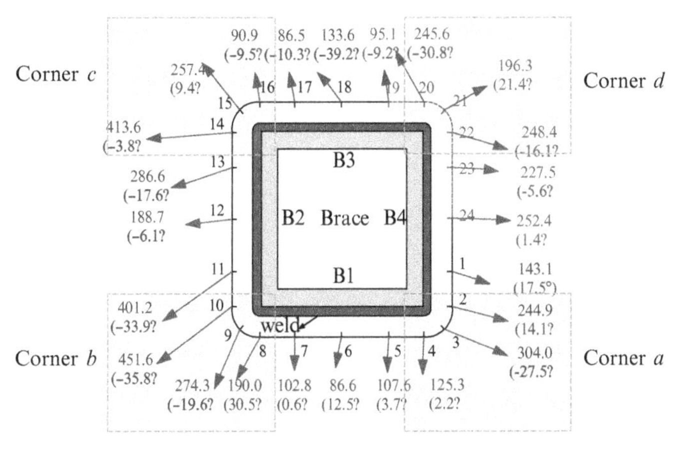

Figure 5.24 The principal stress distribution for the specimen welded at ambient temperature (position: 10/15 mm, unit: MPa).

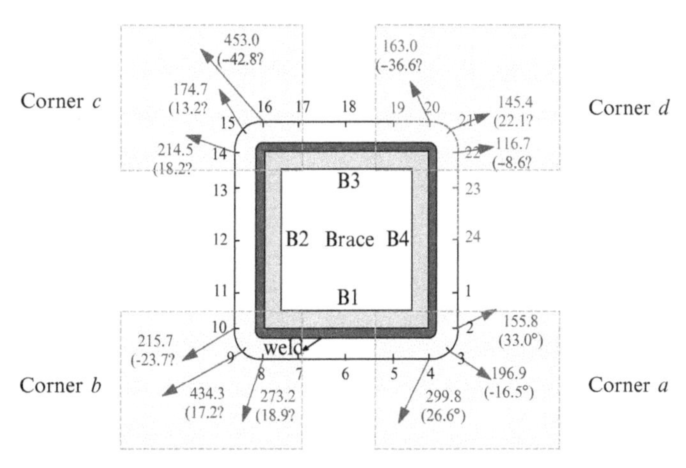

Figure 5.25 The principal stress distribution for the specimen welded at ambient temperature (position: 20/25 mm, unit: MPa).

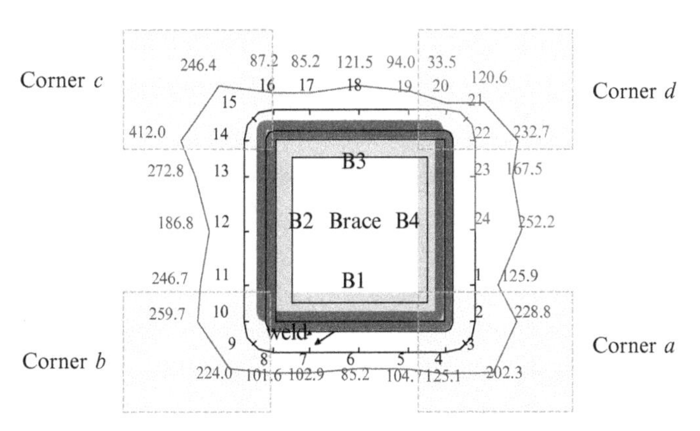

Figure 5.26 The transverse residual stress distribution for the specimen welded at ambient temperature (position: 10/15 mm, unit: MPa).

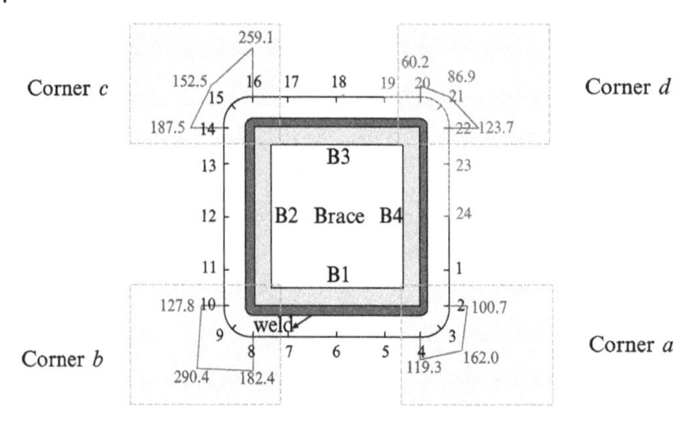

Figure 5.27 The transverse residual stress distribution for the specimen welded at ambient temperature (position: 20/25 mm, unit: MPa).

noticed that the transverse residual stress at 20/25 mm is larger than the value at 10/15 mm for Point 3, Point 8, Point 9, Point 16, and Point 20.

Figures 5.28 and 5.29 show distributions of longitudinal residual stress of the joint welded at ambient temperature. Note that compressive longitudinal stresses turn out at four corners. A noticeable point is Point 9, where the magnitude of longitudinal residual stress is -294.6 (compressive stress). For the other three corners, similar with Point 9, compressive longitudinal residual stress can be found.

Again, the linear interpolation method is used to estimate the transverse and longitudinal residual stress at the chord weld toe and the comparison of both stresses is shown in Figure 5.30. Similar to the preheated specimen, the transverse stress at the chord weld toe is higher than the longitudinal component for a selected point. The tensile transverse residual stress can be found at all points, especially for Point 15, which is located at Corner *c* with a magnitude of 622 MPa. The longitudinal residual stress fluctuates around zero. Figure 5.31 shows the relationship between the residual stresses and the distance from the weld toe. With the increase of the distance, the transverse residual stress shows a decreasing tendency. Table 5.4 lists the residual stresses at measurement points.

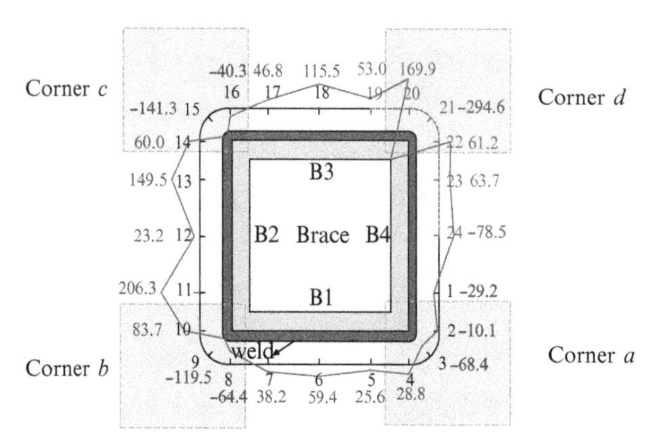

Figure 5.28 The longitudinal residual stress distribution for the specimen welded at ambient temperature (position: 10/15mm, unit: MPa).

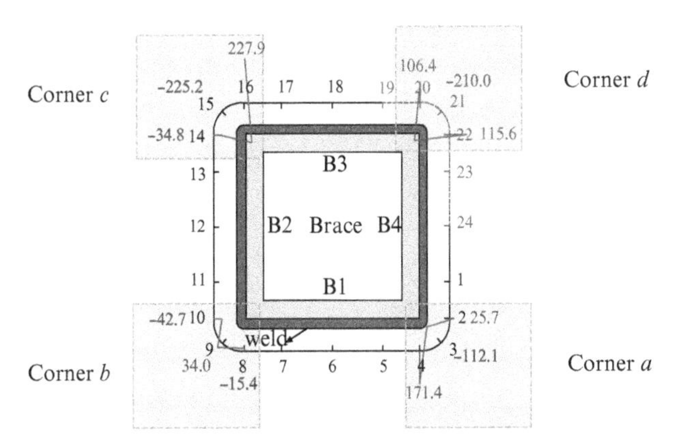

Figure 5.29 The longitudinal residual stress distribution for the specimen welded at ambient temperature (position: 20/25mm, unit: MPa).

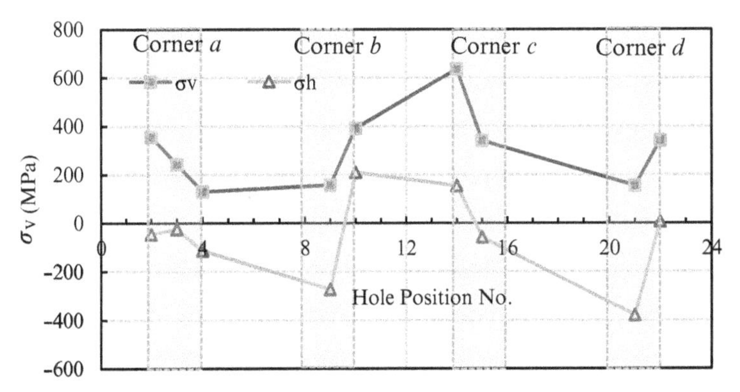

Figure 5.30 The variation of transverse and longitudinal residual stress at the chord weld toe for the specimen with preheating.

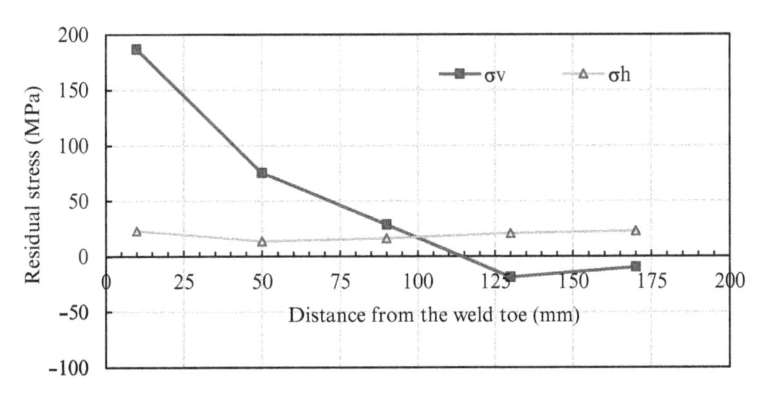

Figure 5.31 The relationship between the transverse and longitudinal residual stress with the distance from the weld toe.

Table 5.4 Residual stress of the specimen welded at ambient temperature.

Hole No.	σ_v (MPa)	σ_h (MPa)	σ_{max} (MPa)	θ (Figure 5.15)
1	125.9	−29.2	143.1	17.5
2	228.8	−10.1	244.9	14.1

(Continued)

Table 5.4 (Continued)

Hole No.	σ_v (MPa)	σ_h (MPa)	σ_{max} (MPa)	θ (Figure 5.15)
3	202.3	-68.4	304.0	-27.6
4	125.1	28.8	125.3	2.2
5	104.7	12.6	113.7	4.8
6	85.2	59.4	86.6	12.5
7	102.9	38.2	102.9	0.6
8	101.6	-64.4	190.0	30.5
9	224.0	-119.5	274.3	-19.7
10	259.7	83.7	451.7	-35.8
11	247.6	206.3	401.2	-33.9
12	186.8	23.2	188.7	-6.1
13	272.8	149.5	286.6	-17.6
14	412.0	60.0	413.6	-3.8
15	246.4	-141.3	257.4	9.4
16	87.2	-40.3	90.9	-9.5
17	83.5	46.8	86.5	-10.3
18	121.5	115.5	133.6	-39.2
19	94.0	53.0	95.1	-9.2
20	33.5	169.9	245.6	-30.8
21	120.6	-294.6	196.3	21.4
22	232.7	61.2	248.4	16.1
23	167.5	23.7	186.3	-12.6
24	252.2	78.5	252.4	1.4

5.4 Analyses and Discussion

5.4.1 Preheating Effect

The comparison of the transverse residual stresses at the chord weld toe between both specimens is given in Figure 5.32. In general, the transverse residual stress at a selected point of the ambient temperature joint is higher than the corresponding value of the preheated joint. This phenomenon is more obvious at the corners (Point 2, Point 10, and Point 14 in Figure 5.32). However, Points 20, 21, and 22 (corner *d*) are the exceptions. Three reasons can explain this phenomenon. One possible reason is uneven preheating. The temperature gradient in this area may still be very high after preheating. Uneven preheating could cause the high temperature gradients in the steel, which makes the cooling rate totally different. In this case, the residual stress in parts of the preheated joint may be higher than the specimen welded at ambient temperature. Secondly, chord edge effect is another significant issue that could change the distribution of residual stress seriously. This will be

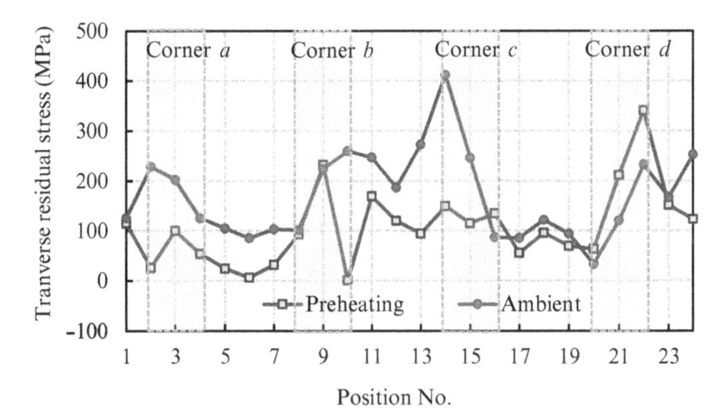

Figure 5.32 Comparison between the transverse residual stresses of two specimens.

discussed in detail in the next section. The third possible reason is the calculation method. It was found in Chapter 4 that the relationship between the residual stress distribution and the distance from the weld toe is highly nonlinear. However, due to experiment conditions (there was not enough space to attach three strain rosettes in the points along B1 and B3 sides, and it was not possible to move the drilling position to the chord weld toe due to equipment limitations), the liner interpolation is used based on the residual stress at 10/15 mm and 20/25 mm to estimate the residual stress at the chord weld toe.

According to Figure 5.32, it can be concluded that preheating can effectively reduce the transverse residual stress. As the finding for the HSS plate-to-plate joints is that when the preheating temperature rises to 300 °C, the residual stress can be eliminated by 80% compared with the case of no preheating treatment, preheating is helpful for residual elimination in box T-joints.

Figure 5.33 shows the comparison for the longitudinal residual stresses in the two joints. Compared with Figure 5.32, the preheating effect is not very obvious for the longitudinal residual stress.

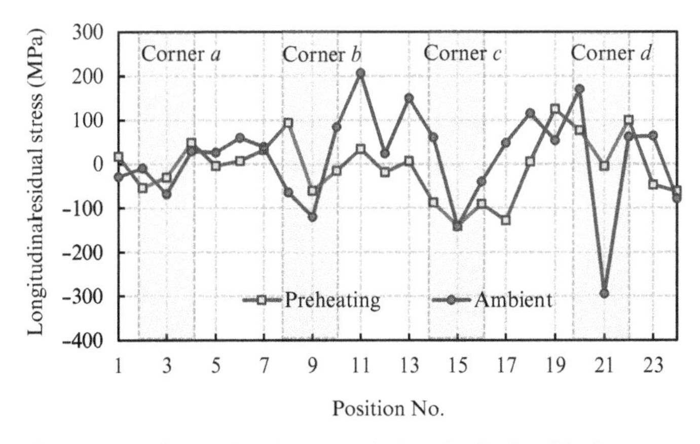

Figure 5.33 Comparison between the longitudinal residual stresses of two specimens.

5.4.2 Chord Edge Effect

It should be noticed from Figures 5.9 and 5.10 that the distance from chord weld toe to the chord box wall is less than 40 mm. Such a short distance causes superposition of the residual stress fields generated during the formation of the chord box and the welding in the intersection part of the joint. This gives an explanation for the phenomenon that for both specimens the residual stress magnitudes at the points along the B1 and B3 sides are generally smaller than the stress values along surface the B2 and B4 side. It also explains the phenomenon that for some points the magnitude of residual stress is higher at 20/25 mm than at 10/15 mm position.

To confirm this, four strain gauges (Point 29, Point 30, Point 31, and Point 32) are added to the specimen welded at ambient temperature to evaluate the stress difference with the points located in the B1 and B3 sides. Figure 5.34 shows the comparison for the residual stress at those points. The transverse residual stresses for Point 25, Point 26, Point 27, and Point 28 are respectively 212.8 MPa, 237.2 MPa, 187.6 MPa, and 194.6 MPa. Compared with the stresses from Point 4 to Point 8 and Point 16 to Point 20, the stresses at Point 25, Point 26, Point 27, and Point 28 are obviously higher. Therefore, part of the residual stress can be eliminated due to the overlay of the residual stress field of chord formation and welding for the joint. Similar results can also be found in the numerical modeling, which will be shown in Chapter 6 in detail.

5.4.3 Corner Effect

It can be found that the welding residual stress distribution around the intersection of the joints is non-uniform. For both specimens, the residual stress magnitudes at the corners are higher than the values tested from at other positions. This corner effect seems more obvious when the measurement positions are 10/15 mm from the weld toe than at 20/25 mm. By taking the preheated specimen for example, in Figure 5.16, the maximum residual stress is located at Point 22, where the principal stress is 582.4

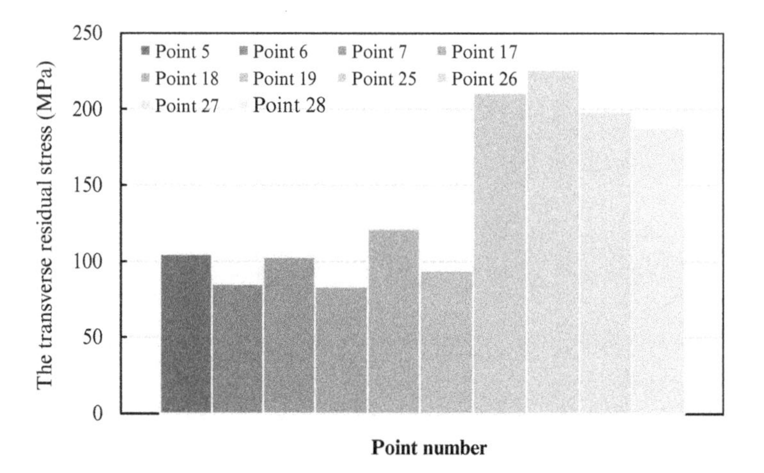

Figure 5.34 Study for the chord edge effect.

MPa, nearly equal to 85% of the nominal yielding stress. In this figure, another impressive location is Point 9, where the principal residual stress is much higher than the nearby points. For the transverse residual stress, a similar conclusion can be obtained from Figure 5.18. However, for the residual stress in the longitudinal direction, as shown in Figure 5.20, this corner effect seems not noticeable. Tensile and compressive residual stress both exist at the joints for the longitudinal component. However, compared with the transverse residual stress, the most evident feature is that the magnitude of longitudinal residual stress is much lower than the transverse residual stress.

Similar findings can also be obtained for the specimen welded at ambient temperature (Figures 5.24, 5.26, and 5.28). A higher residual stress magnitude can be found at the corners than the other positions.

5.4.4 Stress Variation in Depths

In order to find out the residual stress variation in the depth direction at the key locations, Figures 5.35 to 5.38 and Table 5.5 show the relationship between the

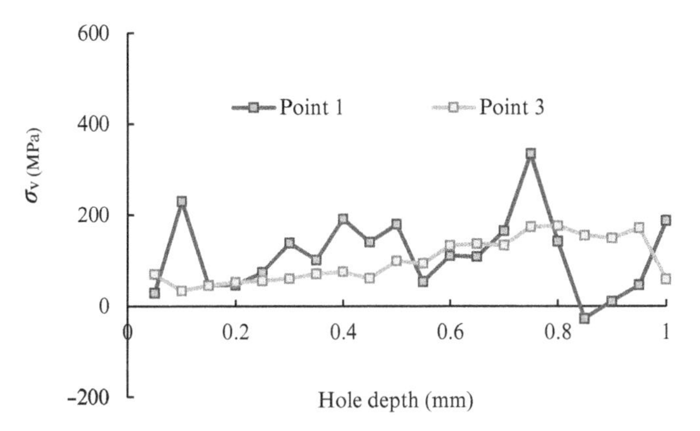

Figure 5.35 The transverse residual stress variation along hole depth (preheated specimen).

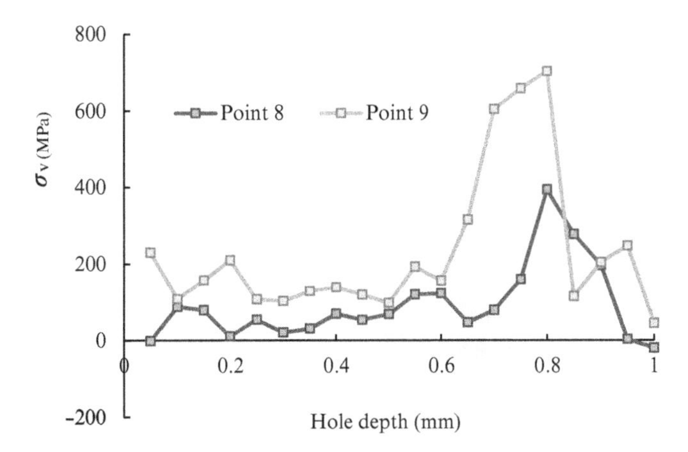

Figure 5.36 The transverse residual stress variation along hole depth (preheated specimen).

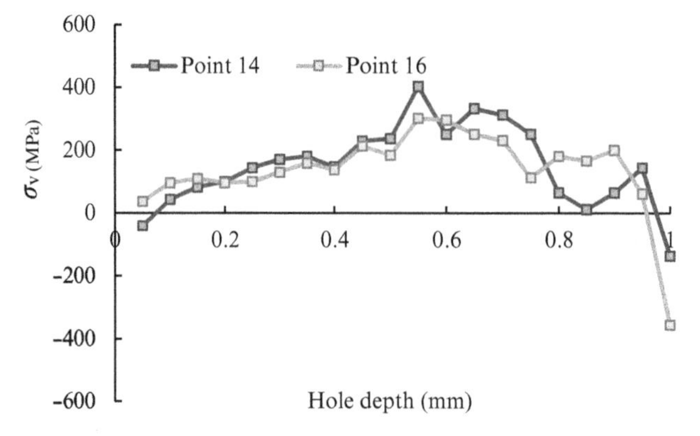

Figure 5.37 The transverse residual stress variation along hole depth (preheated specimen).

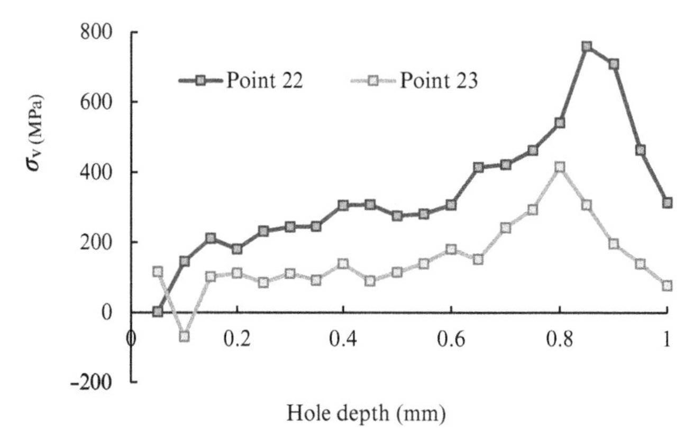

Figure 5.38 The transverse residual stress variation along hole depth (preheated specimen).

transverse residual stress and the depth under the upper surface of the chord for the preheated specimen. Note that the maximum stress is prone to turn out at somewhere between 0.6 mm to 0.8 mm below the upper surface. In this range, part of the material may be yielded due to this in-lock stress. The stress variation from the surface to a depth of 0.4 mm is more moderate compared with the sharp fluctuation in the range from 0.6 mm to 0.8 mm. It is concluded that residual stress distribution along depth is non-uniform.

Table 5.5 Residual stress variation along depth in corners of the specimen with preheating.

Position Hole Depth(mm)	Corner1		Corner2		Corner3		Corner4	
	Point 1	Point 3	Point 8	Point 9	Point 14	Point 16	Point 22	Point 23
0.05	29.1	69.6	−1.3	229.6	-40.8	36.0	1.7	116.5
0.10	230.6	33.5	88.0	108.0	43.6	95.5	145.6	-69.2

Table 5.5 (Continued)

Position Hole Depth(mm)	Corner1		Corner2		Corner3		Corner4	
	Point 1	Point 3	Point 8	Point 9	Point 14	Point 16	Point 22	Point 23
0.15	45.8	45.0	79.4	157.2	82.6	109.6	210.6	102.1
0.20	45.6	52.5	10.5	209.7	99.3	96.4	180.4	112.4
0.25	73.5	55.8	54.6	107.4	144.2	99.2	230.9	85.4
0.30	138.5	60.6	20.6	102.5	170.8	129.8	243.9	111.0
0.35	101.3	70.7	31.2	129.3	180.1	158.8	244.9	92.4
0.40	190.6	75.7	69.7	139.3	146.3	137.9	304.8	138.8
0.45	139.9	61.2	53.9	119.5	229.2	212.5	307.1	89.4
0.50	178.9	98.6	68.7	98.1	236.1	184.1	275.2	114.4
0.55	53.5	93.3	120.0	193.1	402.7	301.1	281.0	139.5
0.60	110.5	132.6	123.2	155.8	250.2	296.3	306.8	180.4
0.65	108.3	136.6	47.1	315.0	332.7	250.8	414.4	151.8
0.70	164.7	133.1	79.2	604.7	312.4	230.8	422.7	242.4
0.75	334.5	173.7	159.7	658.2	250.3	113.2	462.7	293.4
0.80	141.9	175.7	394.0	703.4	65.3	180.4	541.0	416.3
0.85	-27.5	154.9	277.9	115.0	11.3	166.5	758.9	306.4
0.90	10.3	148.8	196.2	204.5	64.5	199.8	709.0	196.1
0.95	46.4	171.1	3.3	247.2	143.7	61.6	464.2	139.3
1.00	187.0	58.8	-19.7	44.9	-136.8	-356.7	313.8	77.7

5.5 Conclusions

Based on the analysis of the test results for both specimens, several conclusions can be drawn. First, preheating is beneficial to reduce the magnitude of residual stress. For both specimens, the residual stress magnitudes on surface B1 and B3 are generally smaller than the stress on surface B2 and B4. In addition, due to the chord edge effect, the magnitude of residual stress is not always higher for the position closer to the weld toe than that at the further position.

6

Numerical Study of Residual Stress for Welded High-Strength Steel Box T/Y-Joints

6.1 Introduction

Fusion welding is a complex process in which the metal parts are heated until they are melted to join materials. Arc physics, thermal transports, microstructure evolutions, and mechanical response are involved in this process. The experimental tools, including hole-drilling method, x-ray diffraction method, and neutron diffraction method, can accurately measure the welding residual stress for structures. However, they are costly and need lots of testing specimens in order to deduce the optimized welding procedure and fabrication procedure. This gives a lot of space for the application of modeling for the study of the residual stress. Since numerical modeling is efficient and cost-saving in the practice, it not only complements the experimental work for the study of welding residual stress with different welding procedures, but also beneficial in designing and operating the manufacturing process. Therefore, many numerical modeling works are suggested for the welding residual stress and the simulation of welding has advanced during past decades. Most analyses are performed with sequentially coupled analysis, which can be divided into two steps, namely the thermal analysis and the mechanical analysis. It is needed to point out that this modeling procedure ignores the effect of heat generation in plastic dissipation. In some cases, it may cause a large and significant change in the fit-up and weld locations. Hence, a fully coupled thermo-mechanical analysis is preferred. When a fully coupled analysis is performed, heat generation caused by mechanical dissipation associated with plastic strain will be included in the analysis. However, the computational cost will be much higher since the mechanical and thermal solutions affect each other strongly and simultaneously. Furthermore, such a procedure requires the use of elements with both temperature and displacement degrees of freedom in the model. To balance the accuracy of modeling and computational cost, some other solutions such as sub-structure model, lumping technique, and adaptive meshing technique are frequently used.

Reducing the dimension of the modeling from 3D to 2D or even 1D is another way to reduce the simulation computational cost. When the deformation or stress in the length direction of a structure can be ignored, the 2D cross section model is frequently used under plane strain conditions. This means the structure and the

Welded High Strength Steel Structures: Welding Effects and Fatigue Performance, First Edition. Jin Jiang.

weld is divided into thin slices perpendicular to the heat source motion direction and the modeling procedure is based on one thin slice. Therefore, though 2D modeling can effectively reduce the computational cost, it also has some disadvantages. For most 2D models, since the analysis is based on a single plane, a reasonable fraction of heat sources is chosen according to measurement or assumption. Therefore, accurate modeling of heat sources is difficult. Furthermore, 2D models can only give the residual stress at selected cross sections and cannot fully describe the residual stress distribution in the whole structure.

In the HSS plate-to-plate joint welding procedure described in Chapter 3, the weld filler is placed in middle of the length of the chord box. Compared with the plate-to-plate joints in Chapters 3 and 4, the welding procedure of box joints is much more complicated. First, different from most traditional rectangular hollow sections, the HSS chord and the brace boxes are formed by welding four steel plates together. This means that the residual stresses have been locked in the chord and brace boxes before welding is carried out at the joint intersection. Second, when the welding is carried out at the joint intersection to connect the chord box and the brace box, superposition of residual stress fields due to welding at the chord and brace boxes and welding at joint intersection complicates the residual stress distribution. Furthermore, different weld profiles and welding parameters such as welding speed, preheating temperature, and welding passes will also affect the residual stress distribution. Besides the welding procedure and welding parameters, the fabrication procedure and joint geometry, joint components also have an influence on the welding residual stress field. When the ratio of the brace box width to the chord box width is changed, the final residual stress will be different. The backing plate may also have an impact on the final residual stress state. Therefore, 3D modeling for the welding residual stress in HSS box joints is helpful to understand the residual stress formed in the welding process. In this chapter, a detailed numerical modeling procedure of residual stress field in HSS box joints is presented. Full thermo-mechanical coupled analysis is used. In order to reduce the computational cost, only the joint intersection part within 0.9 m of the chord box and 0.3 m of the brace box near the chord-brace intersection was selected for analysis. This area is chosen for analysis since it is of most importance for the fatigue performance of the joint under cyclic loads. Two models corresponding to the actual fabricated specimens (100 °C preheated, welded at ambient temperature) were created and validated by testing results. After that, parametric studies were carried out to investigate the impacts of welding parameters such as preheating temperature and weld speeding, as well as geometrical parameters such as the joint skewed angle and ratio of brace to chord on the final residual stress distribution.

6.2 Modeling Procedure

6.2.1 Overview

The main part of the joint near the chord-brace box intersection (chord length: 0.9m, brace length: 0.3m) is extracted for numerical analysis (Figure 6.1). The

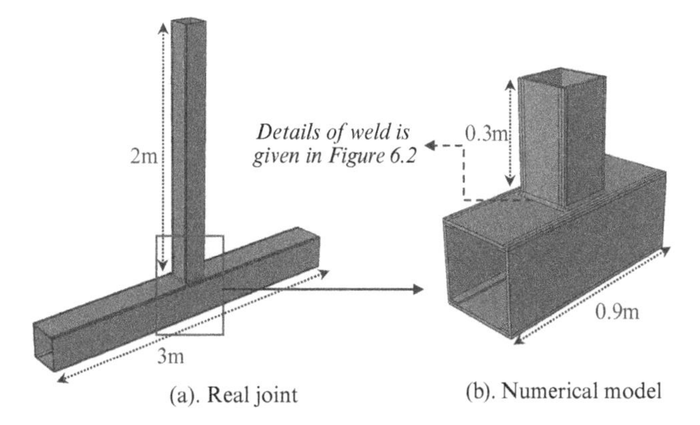

Figure 6.1 Overall model used in the analysis.

modeling procedure of the full thermo-mechanical coupled analysis was divided into two steps. The first step was to simulate the welding residual stress field generated during the box formation. In this step, the multi-pass weld in each corner of the box was combined into only one lump to reduce the computational cost. The second step was to simulate the final residual stress field after the welding was applied at the chord-brace intersection (Figure 6.2). Since the residual stress field in this area is of most importance for the fatigue and fracture performance of the fabricated joint, the weld filler, shown in Figure 5.4, was divided into four lumps. The welding size and the starting welding position in the model followed the actual fabrication procedure (Figure 6.3). The modeling accuracy was validated by comparing the final residual stress field with the testing results.

Two models corresponding to the specimens with preheating treatment and without preheating before the welding, which are viewed as benchmarks models,

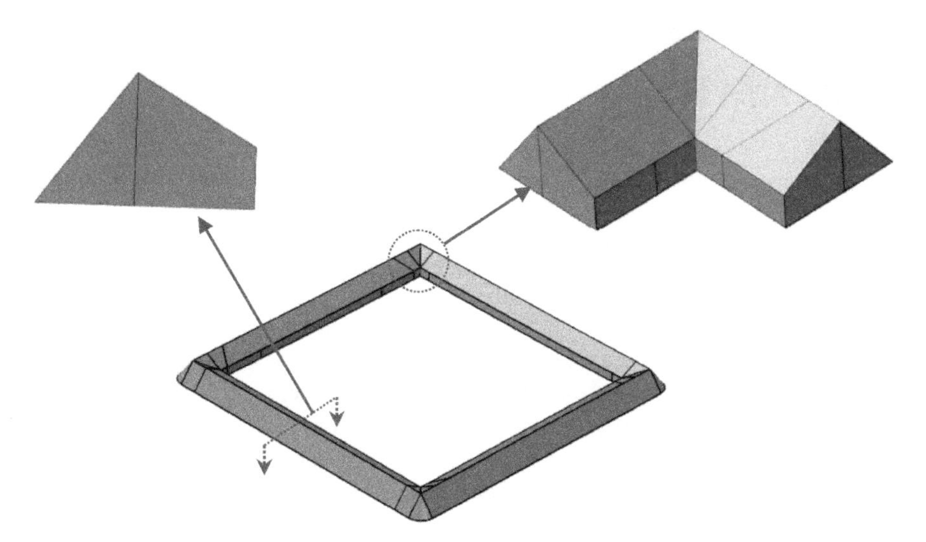

Figure 6.2 Weld near the joint intersection.

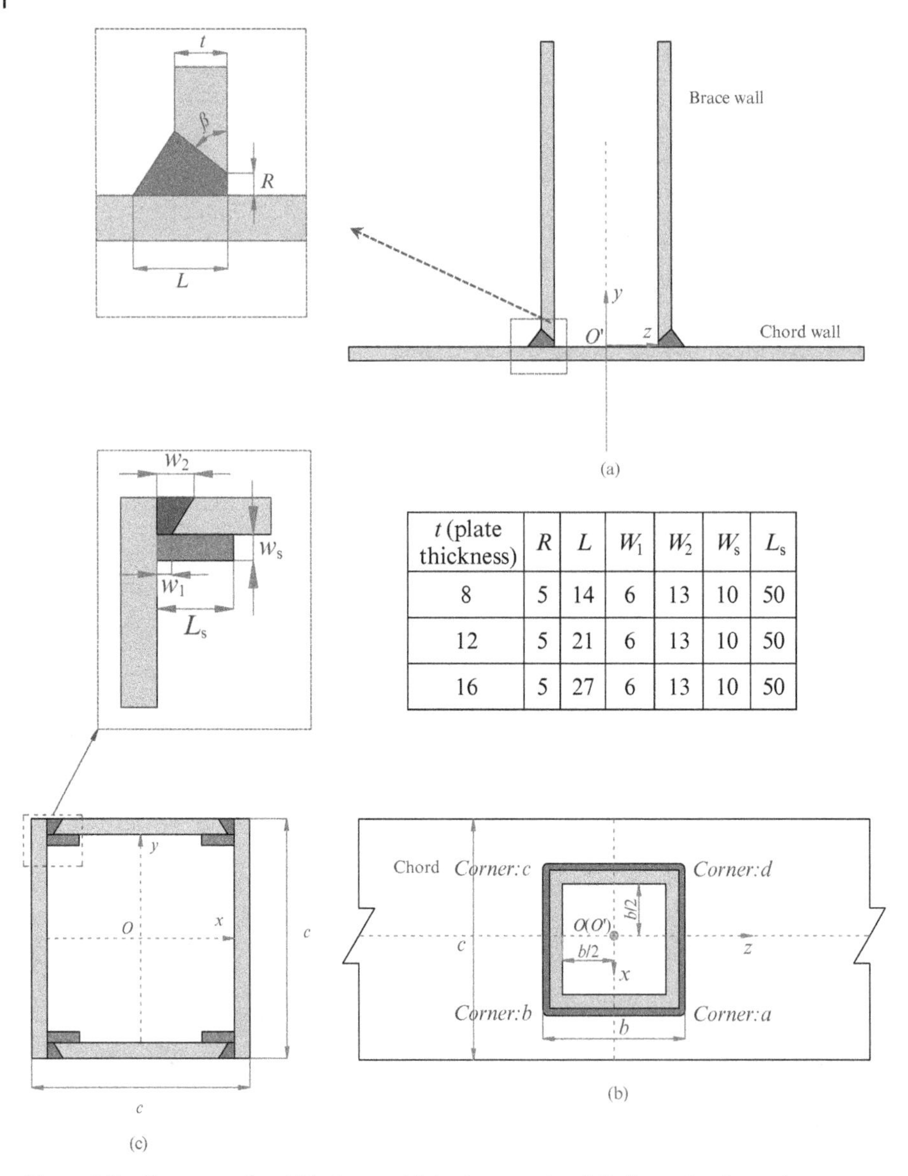

t (plate thickness)	R	L	W_1	W_2	W_s	L_s
8	5	14	6	13	10	50
12	5	21	6	13	10	50
16	5	27	6	13	10	50

Figure 6.3 Geometry of weld in box and joint intersection (all dimension in mm).

were created. For the model corresponding to the preheating, 100 °C was set for the area within 100 mm from the chord weld toe, while 30 °C was set for all other parts. It was input to the model as a predefined temperature field. This means that the preheating temperature error was ignored in the modeling and this predefined temperature field was modeled as uniform in the preheated area. For the model corresponding to the joint without preheating, a consistent temperature field of 30 °C was given to the whole model.

6.2.2 Heat Source Modeling

The accuracy of the heat source model is important for the welding residual stress simulation since the temperature field driven by the heat source is the dominant driving force of the welding process. The transient temperature field outside of the molten weld pool depends primarily on the distribution of the heat source and the conduction properties of the material. The heat source model was used to estimate the heat energy equation, ignoring the complex physics in weld pool. Several generations of the heat source model with increase complexity, from point source model to plane source model to double ellipsoidal model, have been proposed.

In this analysis, the double ellipsoidal model was used to predict the thermal and stress fields during the welding. In the double ellipsoidal model (Figure 4.15), the front half of the source is the quadrant of one ellipsoidal source and the rear half is the quadrant of another ellipsoid. The fractions f_f and f_r ($f_f + f_r = 2$) of the heat deposited in the front and rear quadrants are needed. It is recommended that f_f and f_r can be set to 0.6 and 1.4, respectively (Goldak and Akhlaghi 2005). Since there are four weld clusters at the corners of the box and circumference-going weld filler is added at the intersection of the joint, a FORTRAN program was developed to describe the moving of the heat source in order to ensure that the whole modeling can be completed in a continuous manner (the source code of the program is given in Appendix 3).

6.2.3 Thermal Interactions

In the fusion welding procedure, the weld filler was added gradually into the reserved groove and the thermal interactions between the structure and the air varied with time and space. When the weld filler is added at one corner of the box, heat energy dissipates into the air through convection and radiation on the contact surfaces between the specimen and the air. However, these contact surfaces change with time when the welding is going on. Therefore, in 3D modeling, accurate description of the thermal interaction during the welding process will make the modeling operation very tedious.

In this analysis, a simplified treatment was assumed for the thermal interaction. Different settings for the thermal dissipation surface were given for the residual stress modeling for the box section and for the joint intersection. When the welding is performed in the chord and brace boxes, the free surfaces of the boxes are assumed for thermal convection and radiation. The free surfaces include the surfaces from which the heat can dissipate into the air, except the front free surface in the existing weld (Figure 6.4). There are two main considerations. First is that the ratio of the front free surface in the existing weld to the total free surfaces in the joint is very small (less than 1%). The heat loss in this surface can be ignored. The second reason is that it can save a lot of time in the modeling since the thermal interactions do not need to be adjusted step by step (the position of the front free surface changes with time). On the other hand, when the welding is carried out at the joint intersection part, the case is different. Since the weld length for the joint intersection is much

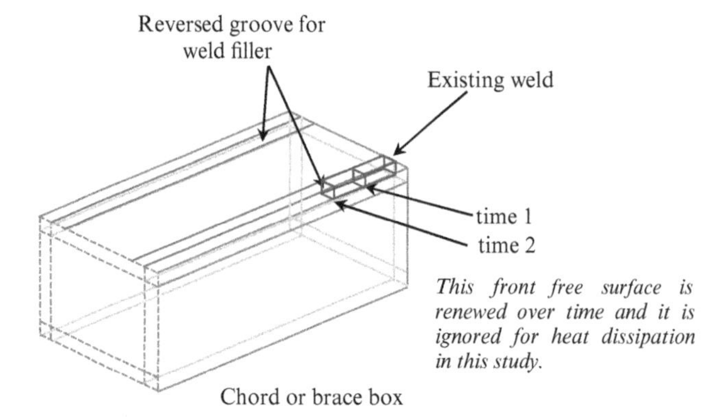

Figure 6.4 Setting for thermal intersection in box formation.

less than the weld of the boxes, the computational cost used in the simulation for residual stress due to the welding at the joint intersection is less than that needed for the box formation. Therefore, the contact surface with air is renewed step by step, including the front free surface in the existing weld.

6.2.4 Arc Touch Movement

As mentioned in Chapter 5, the FCAW method is used in the box joint fabrication, including welding for box hollow sections and welding for chord-brace intersection. It is a semi-automatic arc welding process in which a continuously-fed electrode is used. This welding method is widely used in construction for its high welding speed and portability. However, during the joint fabrication the arc moving speed is controlled by the welder, which causes the welding speed to be non-uniform, especially when the weld is very long and the travelling path is not a straight line. When the arc torch is turned around in the corners of the brace box, the welding speed is frequently reduced to obtain full penetration and good weld quality. In other words, the heat input per length is not uniform in the welding process. Figure 6.5 gives the relationship between the welding speed and the location around the brace box. The average welding speed was 2.8 mm/s, which is obtained by dividing the total weld length by the weld filler adding time. To distinguish the welding speed at different locations, a scale factor $\omega = 1.06$ was applied along the sides of the joint so that the welding speed along the sides of the joint was 3.0 mm/s (1.06×2.8 mm/s), while $\omega = 0.85$ was used at the four corners so that the welding speed at the corners was 2.4 mm/s (0.85×2.8 mm/s).

6.2.5 Modeling Summary

Table 6.1 summarizes and lists the main modeling considerations of the residual stress modeling in the HSS box joints. At the same time, the comparison between the modeling for plate-to-plate joints and box joints is also given in the table. Figure 6.6 gives the detailed operation steps in the modeling.

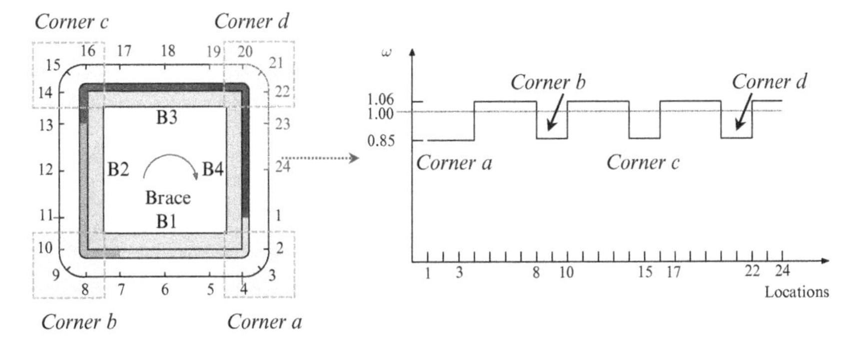

Figure 6.5 Welding speed in chord-brace intersection.

Table 6.1 Comparison of the modeling procedure between plate-to-plate and box joints.

Object of study	Plate-to-plate joint modeling	Box joint modeling
Methodology		
Procedure	Sequentially thermal-stress coupled analysis	Fully coupled thermal-stress analysis
Operation	Heat transfer analysis must first be performed. Stress analysis is performed in which temperatures are applied as external nodal "loads" to generate thermal strain.	Heat transfer and stress analysis are performed together. The impact of plastic strain on the temperature is included.
Element type	Heat transfer element and stress element are respectively selected in thermal and stress analysis	Coupled temperature–displacement element family is selected, which possess both displacement and temperature as degrees of freedom.
Heat model consideration		
Heat input	Heat flux is imported in the model with time-amplitude curve. In the 2D model, the heat flux is added instantaneously in the "slice model" so that a reasonable time range is assumed for heat input.	Double ellipsoidal model is used to predict the thermal and stress field during the welding. Since there are four weld clusters at the corners of the box and circumference-going weld filler added at the intersection of the joint, a FORTRAN program is developed to describe the moving of heat source.
Technique used in the modeling		
Technique	Element birth and death method is used to simulate the process of adding of welding filler. Lumping is used to reduce the computation cost.	All elements are added at the beginning for the welding process of box-forming. However, element "birth and death" is used to simulate the process of adding of welding filler for the chord-brace intersection part. Lumping is used to reduce the computational cost.

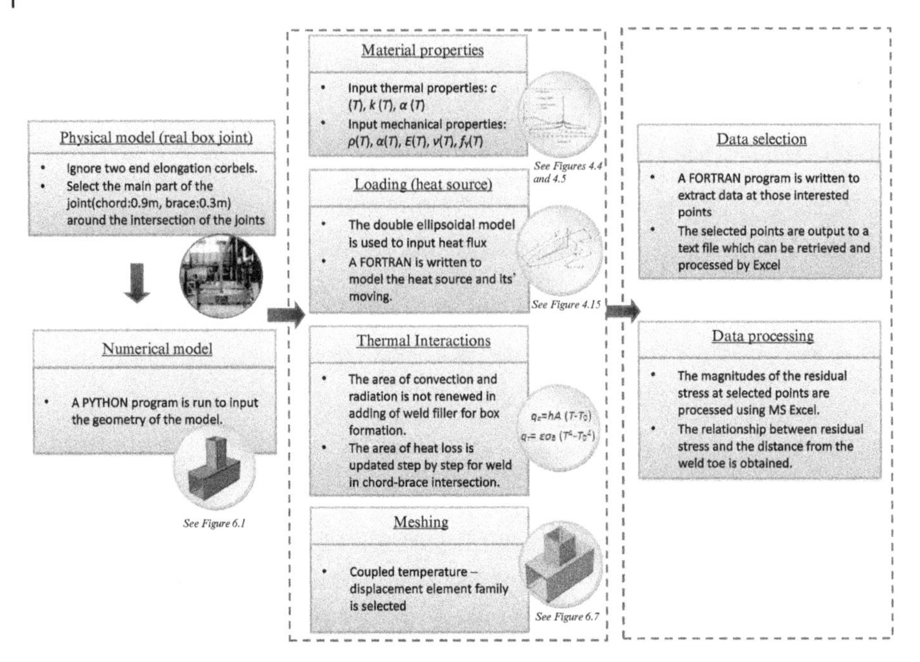

Figure 6.6 Modeling steps for the box joint.

6.3 Modeling of Pure Heat Transfer

In the previous studies, when preheating was carried out, the preheating temperature was simply assigned to the preheated area in the modeling procedure. However, in the actual welding, the temperature field at the moment when weld filler was to be added may not have exactly equalled the designated preheating temperature. This is ascribed to the fact that there is a small time gap (maybe several minutes, depending on the performance of the welder) between the preheating step and the starting of the welding process. Hence, the temperature at the preheated region may have fallen below the preheating temperature when the weld filler is added. In the case when the welding starts minutes later after preheating, the cooling in this time will change the temperature field. Furthermore, it is also very hard to make the preheating temperature uniform in the preheated area. Therefore, it is necessary to evaluate the effect of the time gap from preheating to the starting of the welding on the temperature field of the joint. As a result, a pure heat transfer analysis is given in this section to evaluate the actual temperature at the moment when the weld filler was to be added for the preheated joint.

Figure 6.7 shows the mesh used in the heat transfer analysis. Both elements C3D8R and C3D4R in ABAQUS are selected in the modeling. They are chosen to obtain compatible meshing in the chord box, the brace box, and the weld. Note that in the actual fabrication procedure, the backing plate was spot welded into the box and was not fully attached into the box. Figure 6.8 shows the details of a corner of the box hollow section. Strictly speaking, the areas L_1–L_2 and L_3–L_4 were not fully smeared and there was a very small gap in the adjacent surfaces. Only the area L_2–L_3 is completely welded into the box section. However, in the modeling, the areas L_1–L_2, L_2–L_3, and L_3–L_4 are treated as

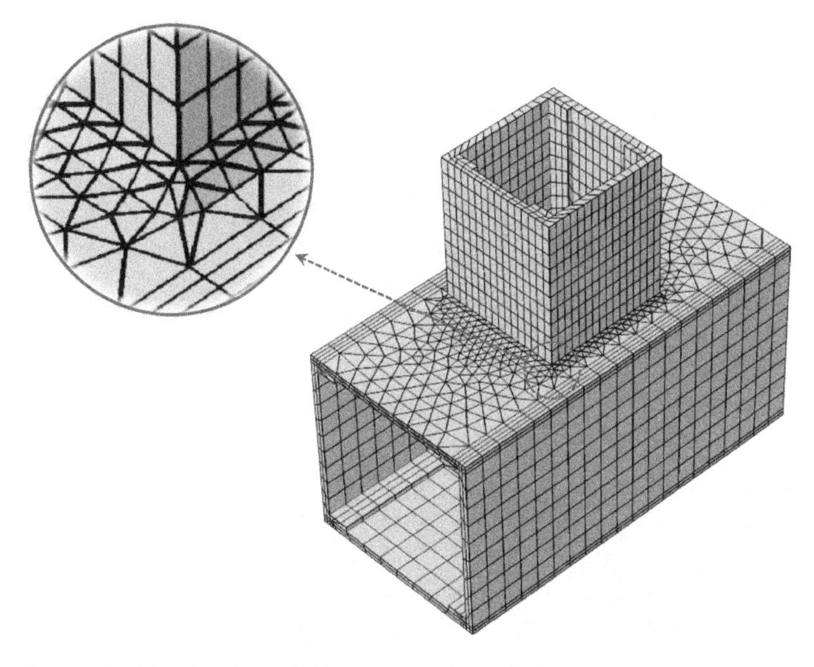

Figure 6.7 Mesh for the model in heat transfer analysis.

fully smeared into the box with the assumption that the gap between the backing plate and the box is zero. To capture the temperature variation along the chord weld toe, 8 monitoring points, corresponding to the numbers in Figure 6.9, were selected for analysis.

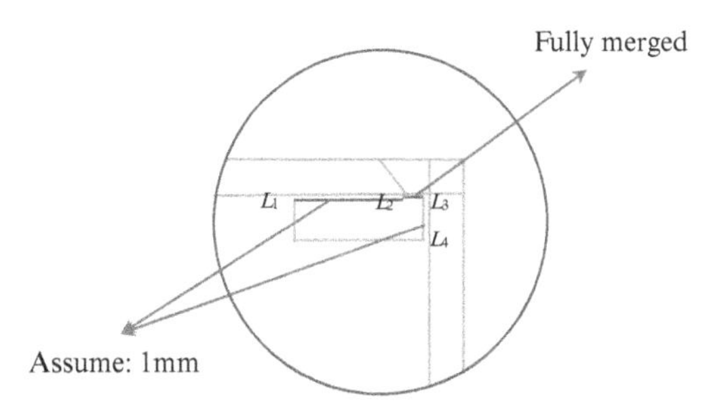

Figure 6.8 Details of a corner of the box section.

In the analysis, a pre-defined temperature of 100 °C was given to the model. The temperature fields at four selected heat propagation times (1 s, 2 mins, 5 mins, and 10 mins) after preheating were selected to show the cooling effect, which is shown in Figure 6.10. Figures 6.10(a) and 6.10(b) respectively give the temperature fields after 1 s and 2 mins of heat propagation. Note that the maximum temperature (located on the corners of the brace box) and the minimum temperature (located in the middle of the plate width of the chord box) are respectively 93.4 °C and 91.1 °C when the propagation time is 2 mins. When the propagation time is 5mins, the

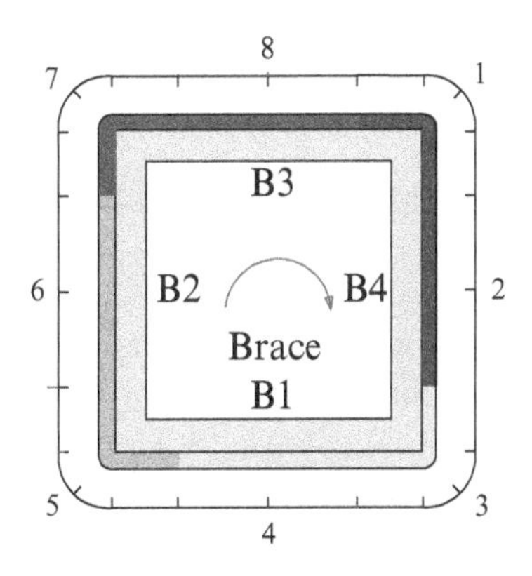

Figure 6.9 Position and numbering of monitoring points.

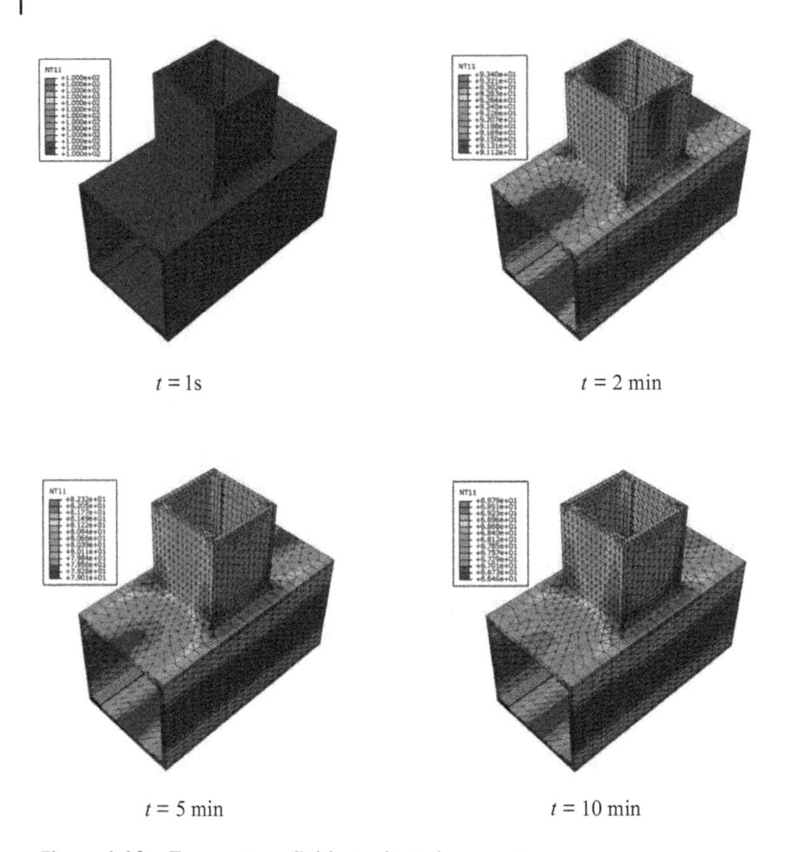

Figure 6.10 Temperature field at selected moments.

maximum and minimum temperature in the joint are respectively equal to 82.3 °C and 79.0 °C. It should be noticed that the locations where the maximum and minimum temperatures are found is the same as when the propagation time is 2 mins. When the propagation time is 10 minutes, the maximum and minimum temperature at the joint is respectively 69.8 °C and 66.4 °C. Therefore, the temperature at different locations after cooling is different and it means that the cooling rate is not constant for different locations of the joint. As shown in Figure 6.10(b), the temperatures at the chord weld toe are higher along the two sides parallel to the chord length direction (B1 and B3 sides shown in Figure 6.9) than the other two sides (B2 and B4 sides shown in Figure 6.9). Similar findings can be obtained from Figures 6.10(c) and 6.10(d). This phenomenon may be ascribed to the fact that the heat is mainly propagated into adjacent chord and brace plates at the B2 and B4 sides while the heat is primarily transmitted into the brace plates at the B1 and B3 sides. The modeling result gives two conclusions: if the time gap between the end of preheating to the start of welding is less than 2 mins, natural cooling in this period does not give much influence on the preheating effect and the initial temperature can be approximated as the preheating temperature. However, if this time gap is large enough (e.g., longer than 5 mins), the cooling during the time gap has obviously influence on the actual preheating effect. Second, the cooling speed is different at different locations of the joint. Generally speaking, the temperature at the chord weld toe along the B2 and B4 sides drops faster than the B1 and B3 sides. To evaluate the temperature dropping speeds at different locations, Figures 6.11 and 6.12 give the temperature distribution and the cooling rates at the 8 selected points when the propagation time is 10 minutes. The locations of these selected points are marked at Figure 6.9. Two series of data, corresponding to the chord weld toe and 15 mm from the chord weld toe, were collected to evaluate the relationship of the temperature and cooling rate with the distance from the weld toe. From the curves, it can be found that the temperature at the weld toe is higher than

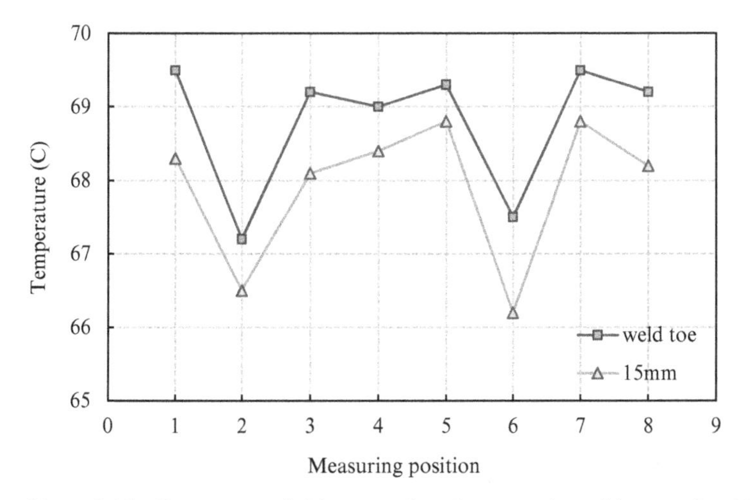

Figure 6.11 Temperature fields comparison between the weld toe and at 15 mm position.

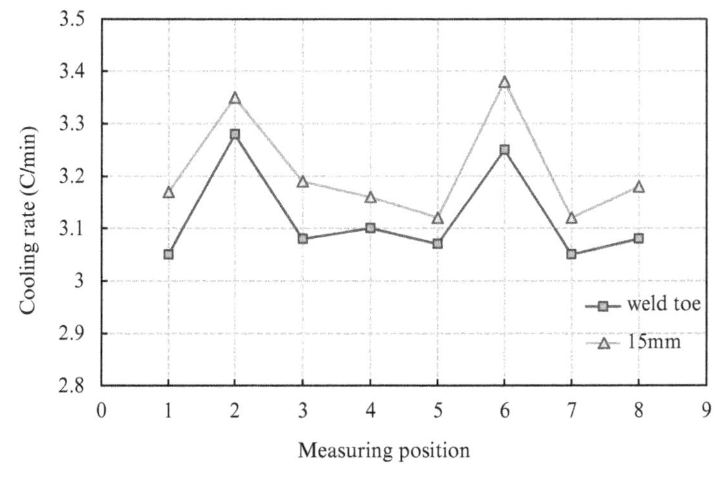

Figure 6.12 Cooling rate comparison between the weld toe and at 15 mm position.

that at 15 mm. This is understandable, since the heat is dissipating from the weld and with an increase of the distance from the weld, the temperature deceases quickly. Two noticeable points are Point 2 and Point 6, which are located in the middle of the chord width and attain lower temperature than the other points.

According to the findings in this section, the initial temperature for the preheated joint can be approximated as the preheating temperature, considering the fact that the welder starts the welding in a very short time (1 mins). In the modeling for residual stress, this short cooling time before the welding for the preheated joint is ignored.

6.4 Fully Coupled Residual Stress Analysis

6.4.1 Modeling Validation

Corresponding to the 24 measurement points around the chord weld toe in the testing, 24 points with the same positioning with experimental investigation were selected to validate the accuracy of the modeling procedure on the residual stress prediction. Since the testing values are obtained at 15 mm from the chord weld toe, the same location (15 mm from the chord toe) was chosen from the modeling results.

Figure 6.13 gives the comparison between the transverse residual stress (perpendicular to the chord weld toe) for modeling and testing results when the joint is preheated to 100 °C. It can be found that when the distance from the chord weld toe is 15 mm, most points agree well for modeling and testing results. Exceptions are Point 10 and Point 22, which are near the corners of the brace box. The difference between the modeling and testing results at Point 10 and Point 22 are respectively 71 MPa and 149 MPa (the difference in term of percentage between the modeling and test results at Point 10 and Point 22 are respectively 27.3% and 64.0%). For other points, the difference between the modeling and testing results is within 50 MPa. Considering the fact that the welder may not keep the torch travelling at a speed constant, especially when it comes to adding weld filler near the brace corners, such results are acceptable. Since a small

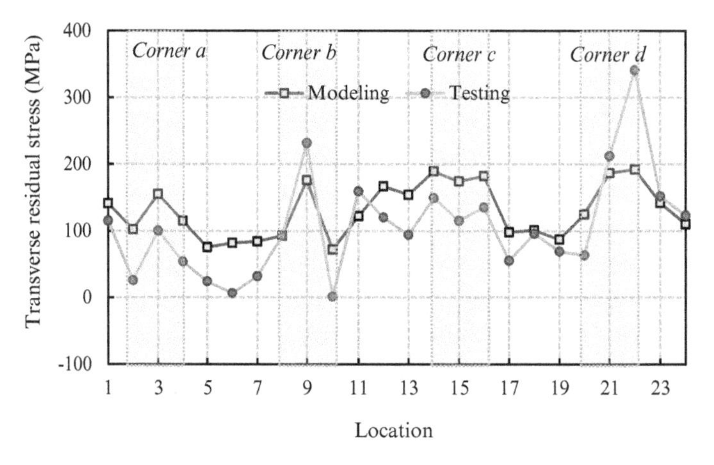

Figure 6.13 Comparison for the transverse residual stress between modeling and testing (preheated).

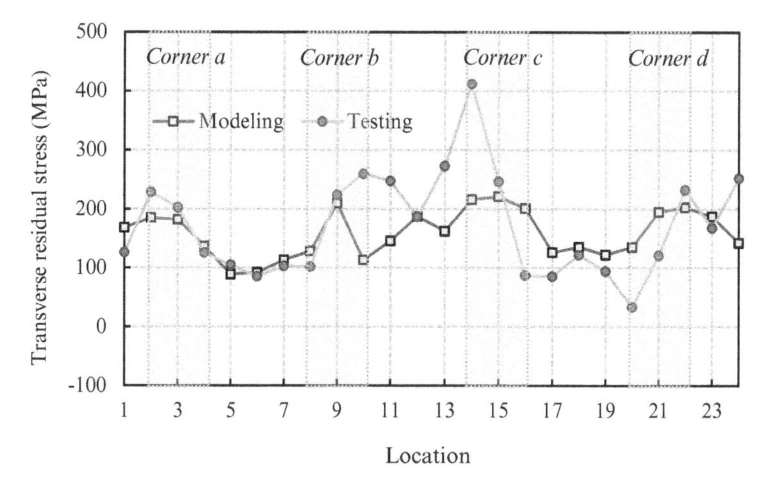

Figure 6.14 Comparison for the transverse residual stress between modeling and testing (ambient temperature).

arc is followed for troch travelling path in the corners, an arc torch is generally moved slower to obtain full penetration weld profile and it means higher heat energy per length is input in the corners than the other parts. The scale factor ω is appropriately estimated based on a time gap to differentiate the welding speed at different locations according to the recorded video in the fabrication yard. However, it is possible that at a certain time, the welding speed set in the modeling at this moment does not agree well with the actual welding speed. Another reason is that the cooling during the period from the end of preheating to the starting of welding may influence the accuracy of the initial temperature field put into the model during the simulation.

Similar results can be found from Figure 6.14, which gives the comparison of the modeling and testing results for the joint welded at ambient temperature. Good agreement between modeling and testing results is obtained at most of points, except for Point 10 and Point 14, which are located near Corner *b* and Corner *c*, respectively (the difference in terms of percentage between the modeling and test results at Point 10 and Point 14 is respectively 58.3% and 49.0%). Unlike the preheated joint, there is no consideration for the influence of the actual initial temperature on the modeling accuracy for the joint welded at ambient temperature. One possible reason for the residual stress gap at Point 14 between modeling and testing result is that the setting for welding speed in the modeling not exactly agreeing with the actual value. But, on the whole, Figures 6.13 and 6.14 give evidence of good agreement for the modeling and testing results at most positions.

6.4.2 Modeling Results

6.4.2.1 Temperature History

Since the temperature fields for the two specimens are similar at the same propagation time, except for the specific magnitude, the result of the model corresponding to the joint welded at ambient temperature was selected to show the temperature fields.

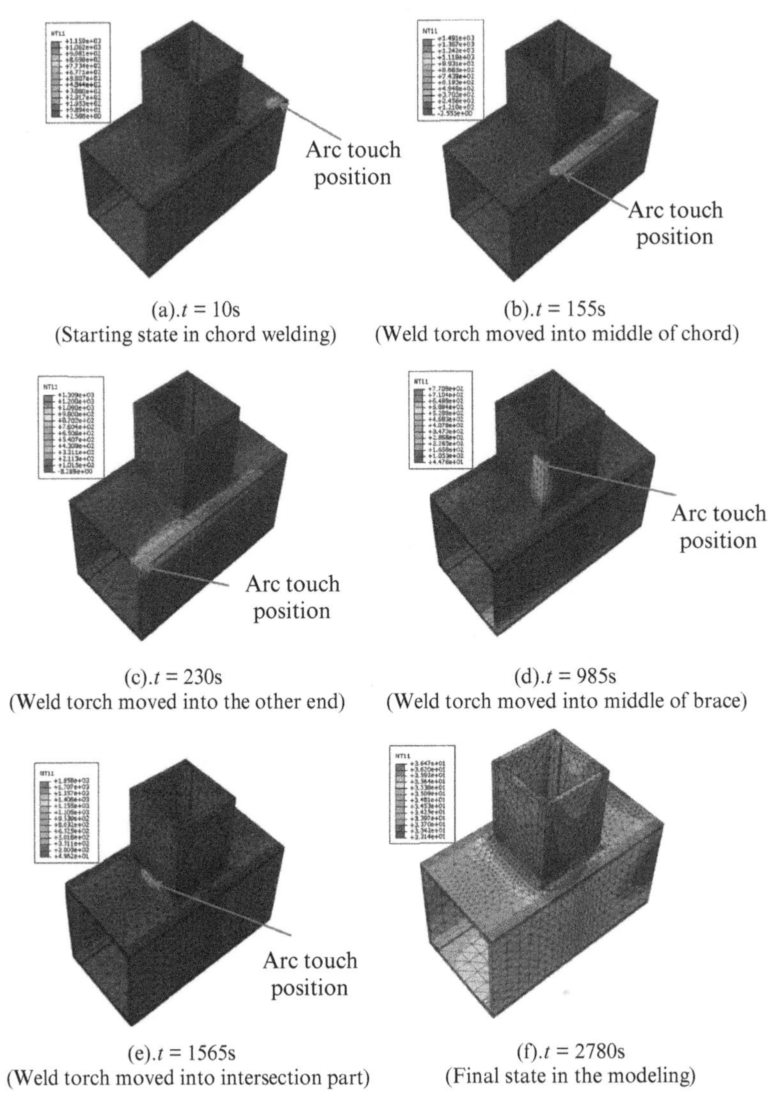

Figure 6.15 Temperature field at selected moments (ambient temperature).

Figure 6.15 shows the travelling path of arc torch during the welding process for the joint welded at ambient temperature. By following the actual welding procedure, the welding torch first moves from one end of the chord box to the other end at the reserved groove along the chord length direction as shown in Figures 6.15(a) to 6.15(c). The maximum temperature of the center of weld pool is around 1800 °C. After that chord is formed with welding applied at four corners, welding torch is shifted into the brace, as shown in Figure 6.15(d). Finally, the heat source is added as shown in Figure 6.15(f).

6.4.2.2 Residual Stress

To show the final residual stress field in the joint, several cross sections are selected to describe the von Mises residual stress state in this section (Figure 6.16). Figure 6.17

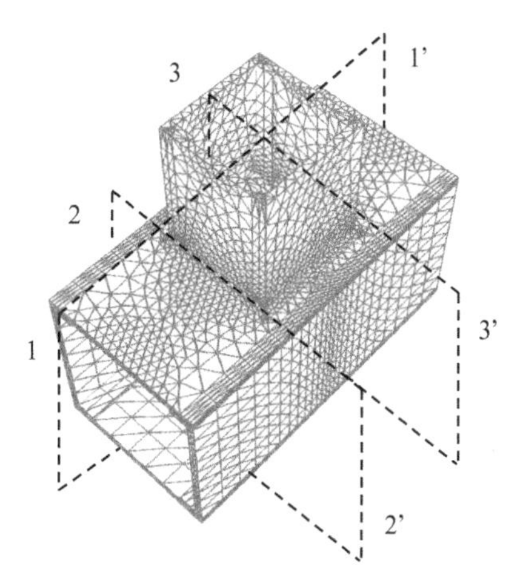

Figure 6.16 Selected cross sections in the joint.

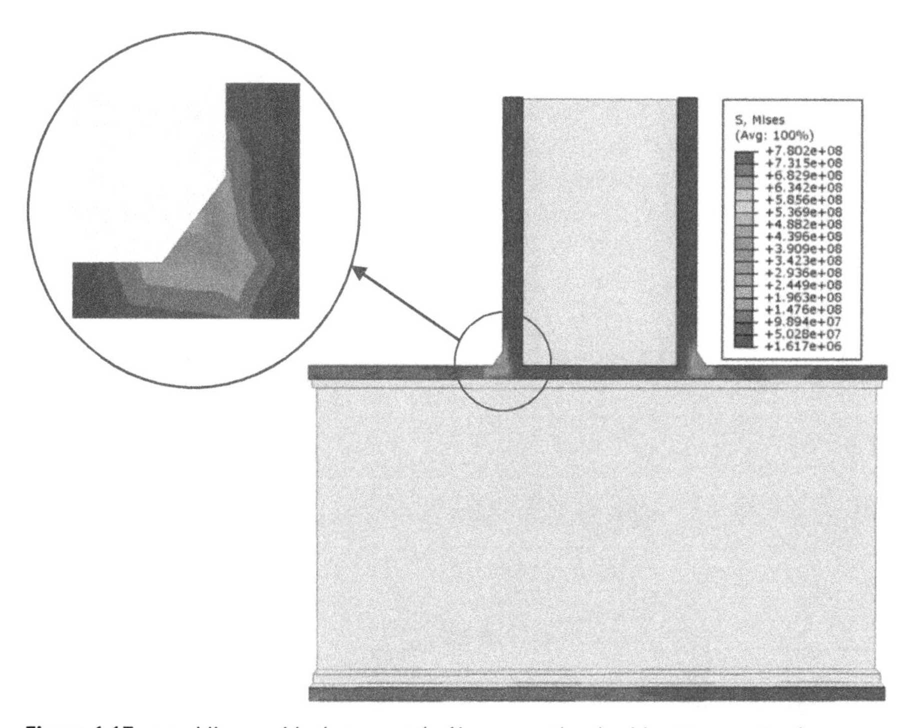

Figure 6.17 von Mises residual stress at 1–1' cross section (ambient temperature).

shows the von Mises residual stress at cross Section 1–1' for the joint welded at ambient temperature when it is completely cooled down to room temperature (Considering that the transverse residual stress is vertical to the chord weld toe and its direction is not constant at different locations around the chord weld toe, the von Mises stress is selected to show the detailed modeling results). Note that the von Mises residual stress in part of the weld is beyond the yielding stress of the material. Another characteristic of the distribution of residual stress that needs to be pointed out is that it is highly non-uniform near the chord weld toe. High residual stress is found within 20 mm of the chord weld toe. When this distance is larger than 20 mm, the residual stress drops quickly. Especially, when this distance is outside 20 mm, the residual stress at the brace is very small (less than 50 MPa). A close-up view at a corner of the box section is given to show the residual stress variation in this area. A layer distribution of residual stress can be found in the area close to the chord weld toe. Higher residual stress is engendered near the top surface of the steel plate than the inner part and bottom surface.

Figure 6.18 shows the von Mises residual stress at cross Section 2–2' for the joint welded at ambient temperature when it is completely cooled down to room temperature. At this cross section, the residual stress at the four corners of the chord box is higher than the other parts of the joint. A layer distribution of the residual stress at the corners can be found at this section. The maximum residual stress turns out at the center of the weld in the corners. Due to the influence of the weld at the chord-brace intersection part, the residual stress at the top surface of the chord box (400–500 Mpa) is obviously higher than the other three steel plates in the chord box (less than 100 Mpa). Another characteristic at the joint is that, in most areas, the residual stress at the top surface is higher than the bottom surface, especially in the middle of the chord width. However, the maximum residual stress is engendered at Corner *b* (the lower left corner in the figure). In other words, for the residual stress at the weld, it is higher at the bottom side. One reason can explain this phenomenon. On the top surface of this cross section, the transverse residual stresses due to the weld at chord box section and due to the weld at chord-brace intersection counteract such that the transverse

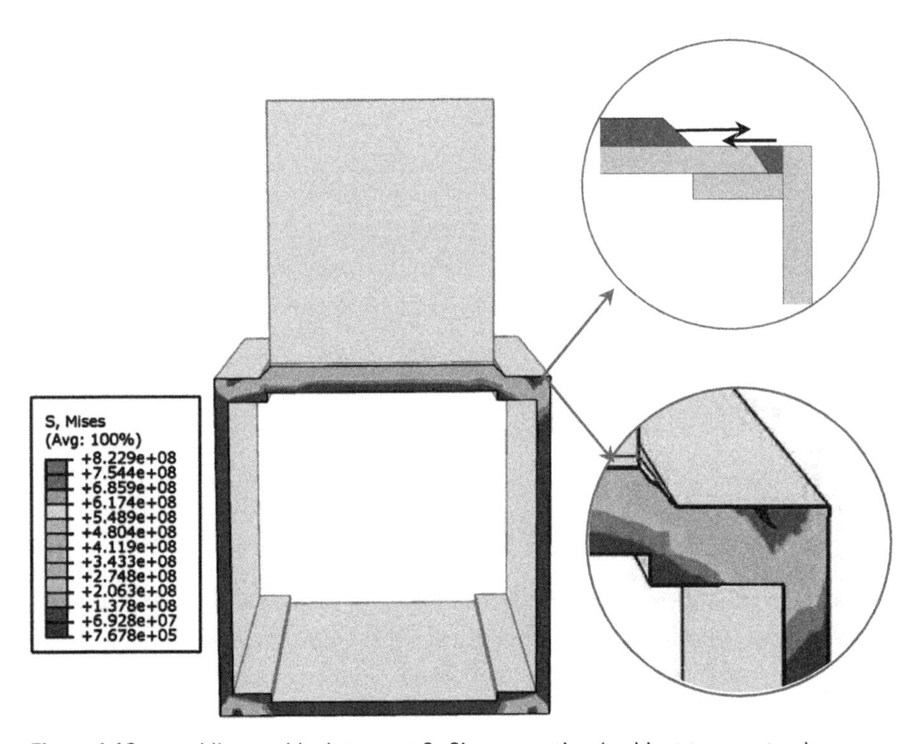

Figure 6.18 von Mises residual stress at 2–2' cross section (ambient temperature).

residual stress is reduced (see close-up view in Figure 6.18). However, the case of the bottom surface of the chord box is different. On the bottom surface of the chord box, only the residual stress due to the weld at the chord box needs be considered (since the weld for the chord-brace intersection is only added on the top surface of the joint and its impact on the bottom surface can be ignored). Figure 6.19 shows the Mises residual stress at cross Section 3–3' of the joint. Unlike Figure 6.18, the maximum residual stress is engendered at Corner *d* (the upper right corner in the figure). Similar residual stress distributions can be found in the joint with preheating.

6.5 Parametric Study

6.5.1 Range of the Modeling

In this section, a parametric study is carried out to find the influence of the welding parameters and joint geometry on the residual stress field for HSS box joints. One hundred sixty-two models were created to investigate the effects of (*i*) the joint angle, (*ii*) the welding starting location, (*iii*) the preheating temperature, (*iv*) the width ratio (b/c), and (*v*) the welding speed on the distributions of final residual stress. The models were run on workstations with Intel Core 2, Quad CPU with a clock speed of 3 GHz, and 8 Gbs of memory. For each model, the computational cost is roughly 30 hours. In order to run the parametric study in an effective way, Python scripting was used in the modeling. Table 6.2 lists the range of the parametric study.

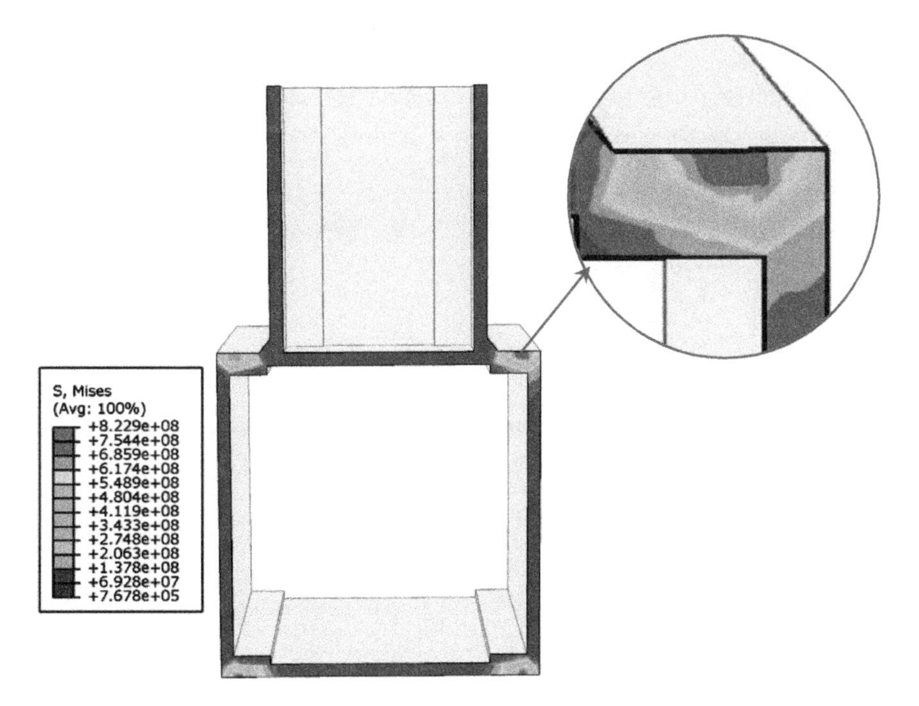

Figure 6.19 von Mises residual stress at 3–3' cross section (ambient temperature).

Table 6.2 Range of the parametric study.

Joint angle (°)	Welding starting location	Preheating temperature (°C)	b/c	Welding speed (mm/s)
90	Middle (Point 16 in Figure 6.20)	30	0.33	2.8
	Close to corner (Point 1 in Figure 6.20)	100	0.50	3.6
		200	0.66	4.2
120	Middle (Point 16 in Figure 6.20)	30	0.33	2.8
	Close to corner (Point 1 in Figure 6.20)	100	0.50	3.6
		200	0.66	4.2
135	Middle (Point 16 in Figure 6.20)	30	0.33	2.8
	Close to corner (Point 1 in Figure 6.20)	100	0.50	3.6
		200	0.66	4.2

Note: *the benchmark models, defined as the models corresponding to the two specimens used in the testing, are (Joint angle: 90°, starting location of welding: Point 24, ambient temperature, b/c: 0.66, welding speed: 2.8 mm/s) and (Joint angle: 90°, starting location of welding: Point 24, preheating temperature: 100 °C, b/c: 0.66, welding speed: 2.8 mm/s)*

6.5.2 Variation of the Residual Stress with Respect to Joint Angle

6.5.2.1 Variation of the Residual Stress with Respect to Joint Angle and Welding Starting Location

In this section, the variation of the residual stress in the HSS box joints (ambient temperature, welding speed: 2.8 mm/s, b/c = 0.67) is studied by varying the parameters of the joint angle (90°, 120°, 135°) and welding start location (near the corner, width middle). Two welding start locations, near the corner (Point 1 in Figure 6.20) and middle of chord box width (Point 24 in Figure 6.20), were selected for analysis. The former one was obtained when the weld filler was added from Point 1, which is the actual welding procedure (Figure 6.21). The latter case was obtained by changing

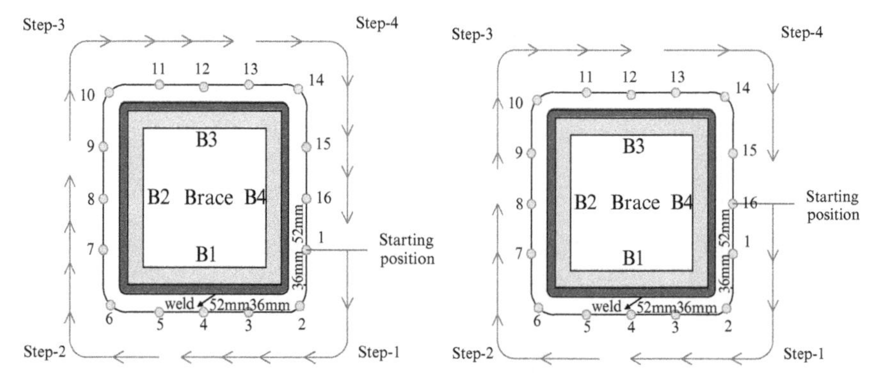

Figure 6.20 Welding start locations selected for analysis.

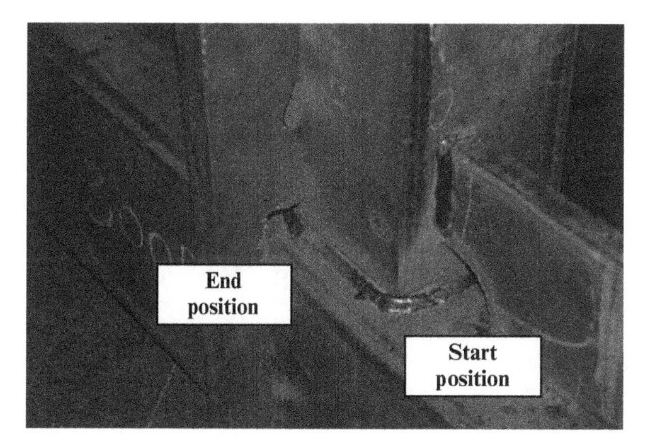

Figure 6.21 The actual welding start position and weld path direction.

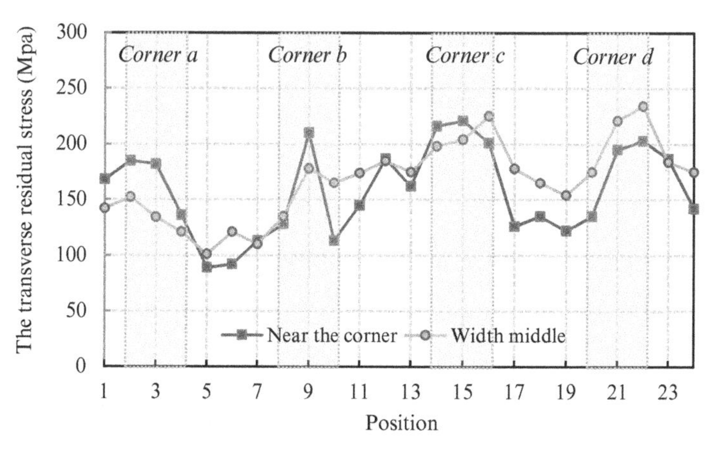

Figure 6.22 Comparison of residual stress with different joint angles (*joint angle: 90°, starting location of welding: Point 1, ambient temperature, b/c: 0.66, welding speed: 2.8 mm/s*).

the welding start location to Point 24 and the same moving direction (clockwise direction) was followed.

Figures 6.22 and 6.23 show the residual stress at selected monitoring points for the 90° and 135° joints, respectively. It can be observed that, for both the 90° and 135° joints, when the weld starting location was chosen at width middle, a lower residual stress can be found from Point 1 to Point 4, which is adjacent to the starting location. However, whatever the weld starting position, there is no obvious difference for the residual stress from Point 5 to Point 24. By comparing Figures 6.22 and 6.23, it can be observed that the residual stress at Point 1 to Point 3 is much higher for the 135° joint when the welding start location is the same.

6.5.2.2 Variation of the Residual Stress with Respect to Joint Angle and Preheating Temperature

Figure 6.24 gives the comparison between the transverse residual stress for three joints welded at ambient temperature with different joint angles (welding start

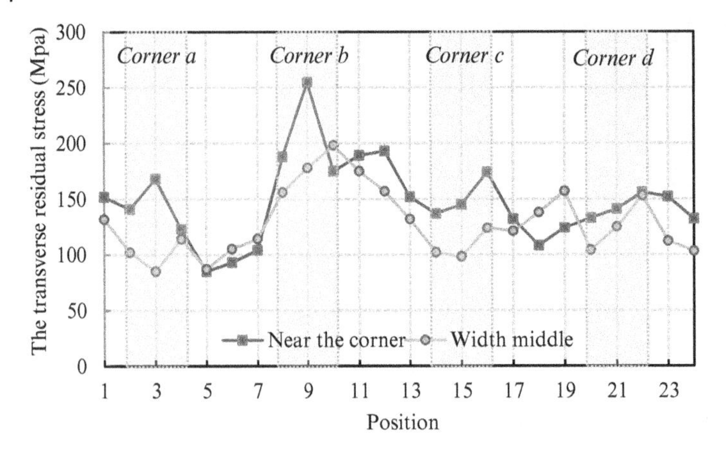

Figure 6.23 Comparison of residual stress with different joint angles (*joint angle: 135°, starting location of welding: Point 1, ambient temperature, b/c: 0.66, welding speed: 2.8mm/s*)

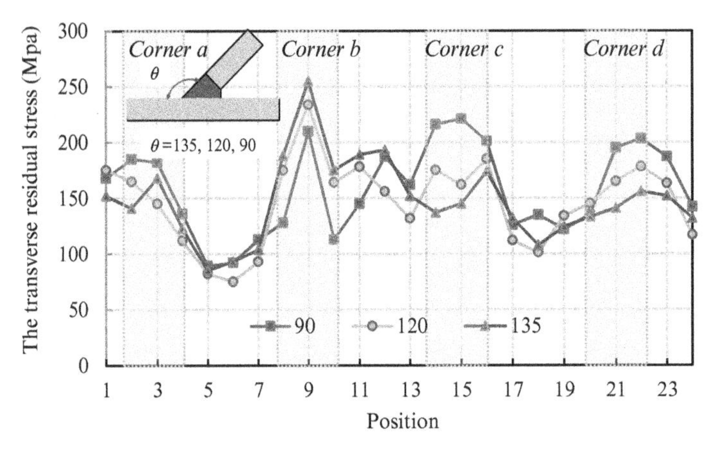

Figure 6.24 Comparison of residual stress with different joint angles (*starting location of welding: Point 1, ambient temperature, b/c: 0.66, welding speed: 2.8 mm/s*).

location: near the corner, welding speed: 2.8 mm/s, b/c: 0.33). It can be observed that higher residual stress turns out at the corners of the chord box, especially at Point 9. From Point 7 to Point 12, higher residual stress is produced at the joint with a larger joint angle. However, from Point 19 to Point 23, the case seems different. In this range, the magnitude of residual stress at 90° is higher than that at the other two angles' joints. The phenomenon can be ascribed to the influence of weld size and cooling rate in the cooling process. When the joint angle is changed, the weld size from Point 8 to Point 16 and Point 20 to Point 4 is correspondingly changed (according to the weld requirement for full penetration for tubular joint from AWS 2008 D1.1).

Figure 6.25 shows the comparison between the transverse residual stress for three welded joints preheated to 100 °C with different joint angles (welding start location: near the corner, welding speed: 2.8 mm/s, b/c: 0.33). Similar to the ambient case shown in Figure 6.24, the residual stress from Point 7 to Point 9 in the 135° joint is

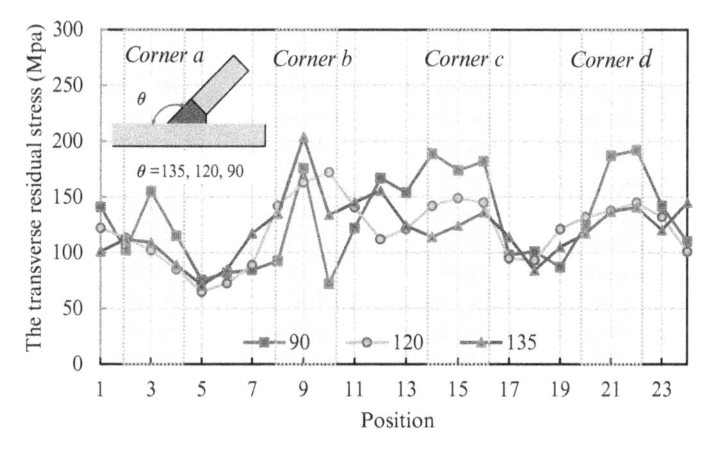

Figure 6.25 Comparison of residual stress with different joint angles (*starting location of welding: Point 1, preheating temperature: 100 °C, b/c: 0.66, welding speed: 2.8 mm/s*).

higher than the 90° and 120° joints. By comparing Figures 6.24 and 6.25, the preheating effect can be observed at a selected point, showing that preheating can effectively reduce part of welding residual stress.

6.5.3 Variation of the Residual Stress with Respect to b/c (Ratio of Brace Width to Chord Width)

6.5.3.1 Variation of the Residual Stress with b/c and Preheating Temperature

Figure 6.26 shows the comparison for the transverse residual stress with different ratios of brace width to chord width for the joints (90°, ambient temperature, welding start location: near the corner). Three values of b/c (0.67, 0.50, and 0.33) are chosen to show the residual stress at selected positions around the chord weld toe. Note that noticeable stress differences can be found at Point 5 to Point 8 and Point 17 to Point 20 for three curves. Picking up two curves corresponding to the cases when b/c is 0.67 and 0.50, it can be found that larger b/c is beneficial to reduce the residual stress at B1 and B3 sides. This means that when b/c is large enough (in other words, the distance between the weld in the chord box and the weld at the chord-brace intersection is close enough, i.e., less than 40 mm), the transverse residual stress between two welds can be reduced due to superposition. However, the residual stress does not show much difference for the cases when b/c is 0.50 and 0.33. This phenomenon demonstrates that this residual stress superposition effect is not obvious when the distance between the two welds is large enough (i.e., larger than 100 mm).

Figure 6.27 gives the comparison for the three cases when the joints (90°, preheated, welding start location: near the corner) are preheated to 100 °C. Similar with Figure 6.26, larger b/c is beneficial to reduce the residual stress at the B1 and B3 sides. For Point 11 to Point 15 and Point 22 to 2, which are located along the B2 and B4 sides, no obvious stress difference can be observed.

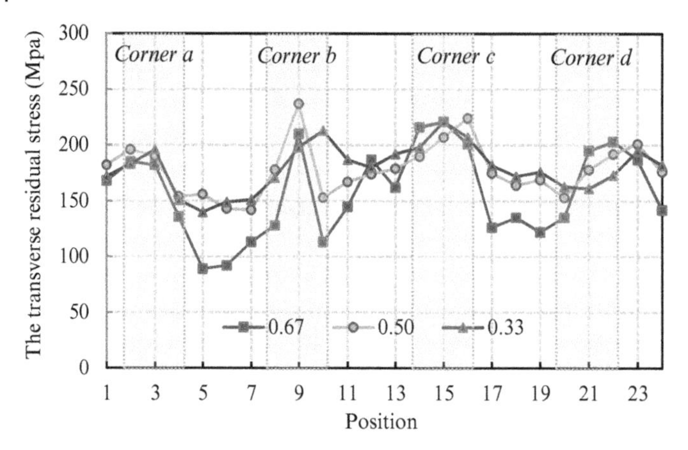

Figure 6.26 Comparison of residual stress with b/c value (*joint angle: 90°, starting location of welding: Point 1, ambient temperature, welding speed: 2.8 mm/s*).

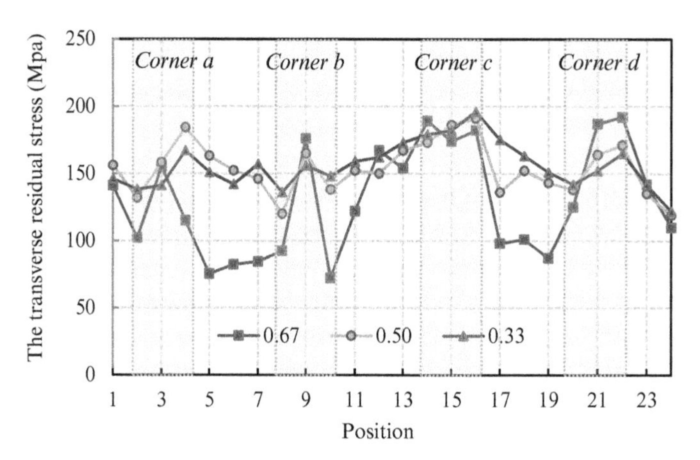

Figure 6.27 Comparison of residual stress with b/c value (*joint angle: 90°, starting location of welding: Point 1, preheating temperature: 100 °C, welding speed: 2.8 mm/s*).

6.5.3.2 Variation of the Residual Stress with b/c and Welding Starting Location

Figure 6.28 gives the comparison for the transverse residual stress with different welding start locations for the joint (b/c = 0.50, 90°, welded at ambient temperature). Two cases, corresponding to the cases when the welding start locations are chosen near the chord box corner (Point 1 in Figure 6.9) and the middle of the chord box width (Point 24 in Figure 6.9), were selected for analysis. Note that a slight difference can be found at Point 1 to Point 3 for two curves. It can be found that when the welding start location was chosen to be at middle of chord width, the residual stress at Point 1 to Point 3 can be slightly reduced when compared with the residual stress obtained. For the other positions, there is no obvious difference between two curves.

A similar conclusion can also be drawn from Figure 6.29, which shows the comparison of the transverse residual stress for the joints (b/c = 0.33, 90°, welded at

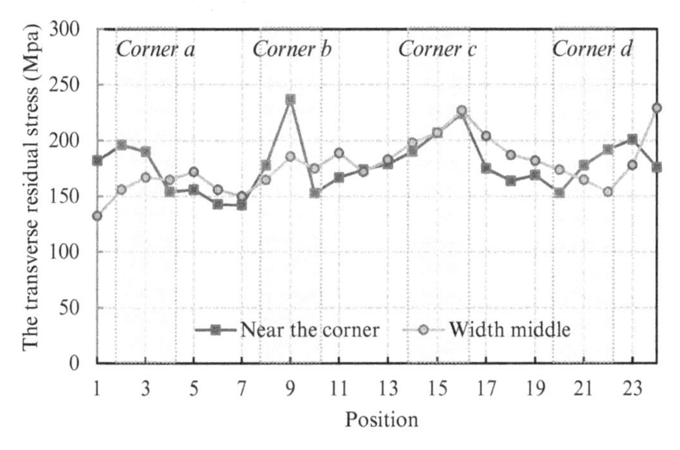

Figure 6.28 Comparison of residual stress with different welding start location (*joint angle: 90°, preheating temperature: 100 °C, b/c: 0.50, welding speed: 2.8 mm/s*).

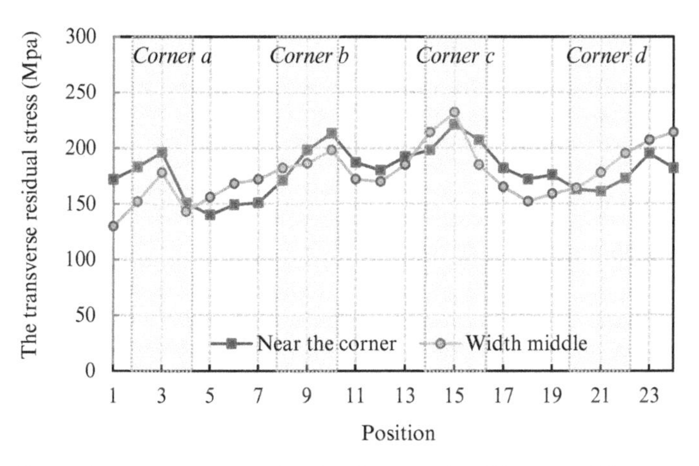

Figure 6.29 Comparison of residual stress with different welding start location (*joint angle: 90°, preheating temperature: 100 °C, b/c: 0.33, welding speed: 2.8 mm/s*).

ambient temperature). In Figure 6.29, at Point 1 to Point 3, a higher residual stress can be found at the case when the welding start location was chosen to be at near the corner.

6.5.4 Variation of the Residual Stress with Respect to Welding Speed

6.5.4.1 Variation of the Residual Stress with Welding Speed and Preheating Temperature

Figure 6.30 gives the comparison of transverse residual stress around the chord weld toe at different welding speeds for the joint (90°, welded at ambient temperature, b/c = 0.67). Three welding speeds (2.8 mm/s, 3.6 mm/s, and 4.2 mm/s) were selected to evaluate its impact on the final transverse residual stress field around the

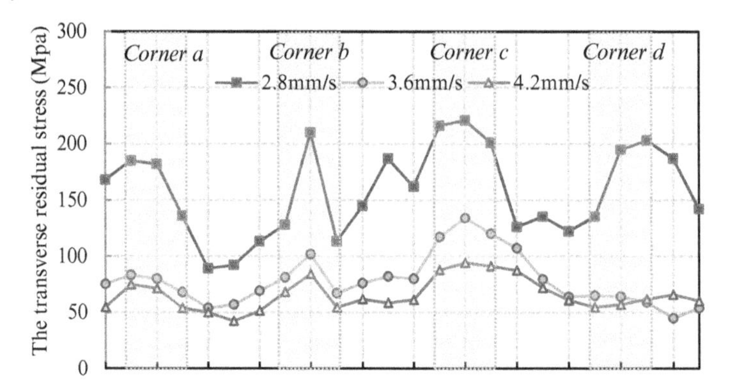

Figure 6.30 Comparison of residual stress with different welding speeds (*joint angle: 90°, ambient temperature, b/c: 0.67, welding starting location: Point 1*).

chord weld toe. It can be observed from Figure 6.30 that that higher welding speed can effectively reduce the magnitude of residual stress. Especially, when the welding speed is increased from 2.8 mm/s to 3.6 mm/s, the reduction of residual stress at Point 9 is larger than 100 MPa. Note that when the welding speed is increased from 3.6 mm/s to 4.2 mm/s, a small decrease of the transverse residual stress can be observed at most of the points, though this decrease magnitude is much smaller than when the welding speed is reduced to 3.6 mm/s from 2.8 mm/s. A similar conclusion can also be found from Figure 6.31, which shows the comparison of transverse residual stress for the joint (90°, preheating, b/c = 0.67).

It is concluded that a higher weld speed can reduce the magnitude of residual stress. With an increase of welding speed, the linear heat input is reduced in proportion when the arc voltage and current are constant. Therefore, an increase of welding speed can effectively reduce welding residual stress and can also improve the fabrication efficiency at the same time. However, if the weld speed is too fast, it will possibly bring some defects such as gas holes and incomplete penetration. Generally, in order to reduce the sensibility of cracking and improve the mechanical properties of the weld, a minimum penetration ratio should be satisfied. On the other hand, when the torch travelling

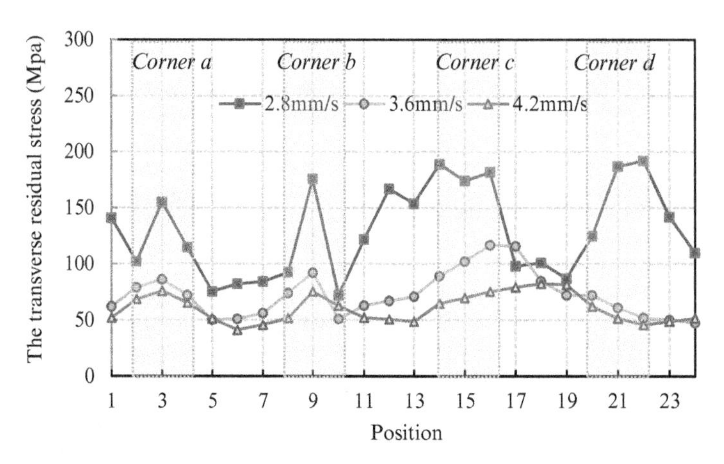

Figure 6.31 Comparison of residual stress with different welding speeds (*joint angle: 90°, preheating temperature: 100 °C, b/c: 0.67, welding starting location: Point 1*).

speed is too slow, beside an increase in the magnitude of residual stress, the area of HAZ will be increased and crystal grain will become larger. Therefore, a reasonable range of welding speed is needed during the welding procedure.

6.5.4.2 Variation of the Residual Stress with Welding Speed and b/c

Figure 6.32 gives the comparison of transverse residual stress around the chord weld toe at different welding speeds for the joint (90°, welded at ambient temperature, b/c = 0.50). Three welding speeds (2.8 mm/s, 3.6 mm/s, and 4.2 mm/s) were selected to evaluate the impact on the final transverse residual stress field around the chord weld toe. For all the curves, higher residual stresses can be found at the corners. As shown in Figure 6.30 and Figure 6.31, at a selected point the residual stress for the case when average welding speed is 2.8 mm/s is much larger than the stresses when the average welding speed is 3.6 mm/s and 4.2 mm/s.

Figure 6.33 gives the comparison of transverse residual stress around the chord weld toe at different welding speeds for the joint (90°, welded at ambient temperature,

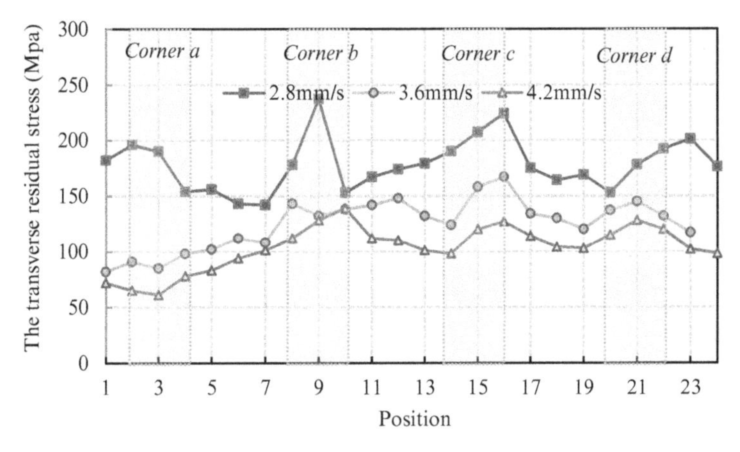

Figure 6.32 Comparison of residual stress with different welding speeds (*joint angle: 90°, preheating temperature: 100 °C, b/c: 0.50, welding starting location: Point 1*).

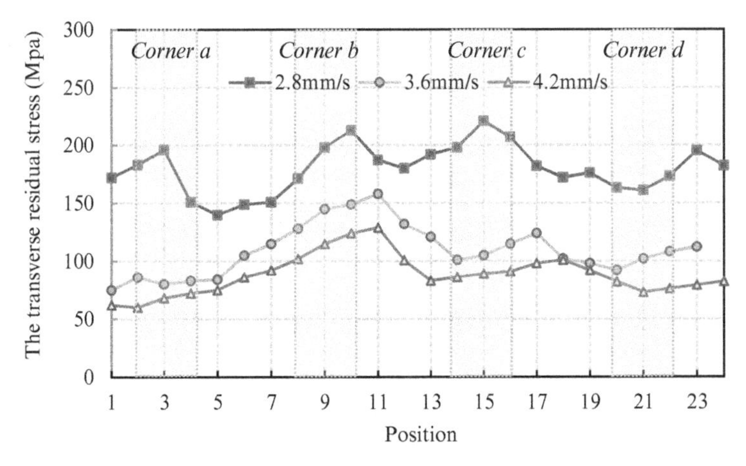

Figure 6.33 Comparison of residual stress with different welding speeds (*joint angle: 90°, preheating temperature: 100 °C, b/c: 0.33, welding starting location: Point 1*).

b/c = 0.33). Similarly, three welding speeds (2.8 mm/s, 3.6 mm/s, and 4.2 mm/s) were selected to evaluate the impact on the final transverse residual stress field around the chord weld toe. From Figure 6.33, it can be observed that when b/c is equal to 0.33, higher welding speed can give lower welding residual stress for all positions around the chord weld toe.

6.6 Conclusions

This chapter presents the 3D procedures of fully coupled thermo-mechanical analysis for a set of HSS box welded joints. A modeling procedure for the simulation of welding residual stress for welded box joint is proposed. Good agreements are obtained for two benchmark models, corresponding to the specimens used in Chapter 5. It is found that the fully coupled thermo-mechanical analysis is an efficient way to model the welding residual stress. The lumping technique and the element birth and death technique are useful in balancing the computational cost and accuracy of the modeling.

By heat transfer analysis, it is found that the temperature at different locations after cooling is different, which means the cooling rate is not constant for different locations of the joint. After cooling, the temperatures at the chord weld toe are higher along the two sides parallel to the chord length direction (B1 and B3 sides shown in Figure 6.9) than the other two sides (B2 and B4 sides shown in Figure 6.9). If the time gap between the end of preheating to the start of welding is less than 2 mins, the natural cooling in this period does not give much influence on the preheating effect and the initial temperature can be approximated as the preheating temperature. However, if this time gap is large enough (e.g., longer than 5 mins), the cooling during the time gap has an obvious influence on the actual preheating effect. Second, the cooling speed is different at different locations of the joint. Generally speaking, the temperature at the chord weld toe along the B2 and B4 slides drops faster than the B1 and B3 sides. In addition, it is found that the temperature at the weld toe is higher than at 15 mm at the same propagation time.

In this chapter, a small-scale parametric study was carried out to evaluate the effects of the joint angle, welding start location, preheating temperature, b/c, and the welding speed on the distributions of final residual stress. It is found that the preheating treatment can reduce the magnitude of the principal residual stress at the weld toe. It can be observed that, when the weld starting location is chosen at the middle of the brace width, a lower residual stress can be found from the area adjacent to the starting location. However, whatever the weld starting position is, there is no obvious difference for the residual stress at the welding starting position. When b/c is large enough (in other words, the distance between the weld in the chord box and the weld at the chord-brace intersection is close enough, i.e., less than 40 mm), the transverse residual stress between two welds can be reduced due to superposition. However, the residual stress does not show much difference for the cases when b/c is 0.50 and 0.33. This residual stress superposition effect is not obvious when the distance between the two welds is large enough (i.e., larger than 100 mm). In addition, the welding speed has significant influence on the magnitude of residual stress at the weld toe. Higher speeds can effectively relieve the residual stress at the weld toe.

7

Stress Concentration Factor of Welded Box High-Strength Steel T-Joint

7.1 Introduction

The fatigue performance of tubular joints under cyclic loads is one of the major concerns in the design of tubular structures. Many studies can be found for the fatigue investigation of the welded joints made of mild steel. Two main categories can be summarized, namely the S–N approach and the fracture mechanics method. The S–N approach is based on a simple relationship between applied stress range ΔS and the number of cycles N to failure. When a constant amplitude loading is applied, the S–N curves are assumed to be linear for a log-log scale until the fatigue limit is reached. The merits of the S–N approach are that it is simple to use and fatigue life can be evaluated by checking the geometry, loading case, and environment for the joint. When a joint has a crack-like default, the S–N curves are no longer applicable, whereas the fracture mechanics method can be used to describe the crack behavior. The basis of the fracture mechanics method is that it considers the stress field and not just the stress concentration at the weld toe.

Up to now, plenty of large-scale fatigue tests on hollow section joints have been reported. However, most of those studies aimed to investigate the fatigue life of CHS joints and RHS joints made of mild steel. The understanding of the fatigue performance of the HSS BHS joint is still a vacancy. Although some equations about the SCF exist under axial load (AX) and in-plane bending (IPB), the SCF under out-of-plane bending (OPB) is not ready yet, limiting the application of the superposition method in calculating the hot-spot stress under a combined load. Furthermore, the uncertainty brought by the welding residual stress in the BHS fabrication for the fatigue performance is unclear.

Based on the current situation of the BHS, a static test on large-scale preheated BHS T-joints made of HSS RQT701 was conducted in this study. The objective of the test was to study the SCF distributions under static loading. In the static test, the SCF distributions under three basic loads (AX, IPB, and OPB) were obtained and the SCF values were used to verify the existing equation and the superposition method. In addition, the impact of welding residual stress on the stress concentration factor is also evaluated in this chapter.

Welded High Strength Steel Structures: Welding Effects and Fatigue Performance, First Edition. Jin Jiang.

7.2 Test Setup and Specimens

The orange test rig designed for uni-planar hollow section joint under AX, IPB, OPB, or any combinations of these three basic load was used for static and fatigue test for the BHS joints (Figure 7.1). The MTS series 244 hydraulic actuators were equipped in the test rig. These are double-acting actuators that operate under a precision servo-valve control in a servo-hydraulic system. Actuator piston rod movement is accomplished by supplying high-pressure hydraulic fluid and the differential pressure across the piston forces the piston rod to move. The internally mounted linear variable differential transformer (LVDT) provided an accurate indication of the move displacement of the piston rod.

The orange test rig is capable of capturing the SCF and hot-spot stress around the weld toe by applying static loads. In addition, it can measure the fatigue life of the joint by applying cycle loads. Three actuators, actuator AX, actuator IPB, and actuator OPB, were equipped in the orange test rig to apply loads along three mutually perpendicular axes (Figure 7.2). The maximum load that can be applied for actuator AX and actuator IPB is 250KN. The load capacity of actuator OPB is 150KN. The actuators can be operated individually or concurrently to create

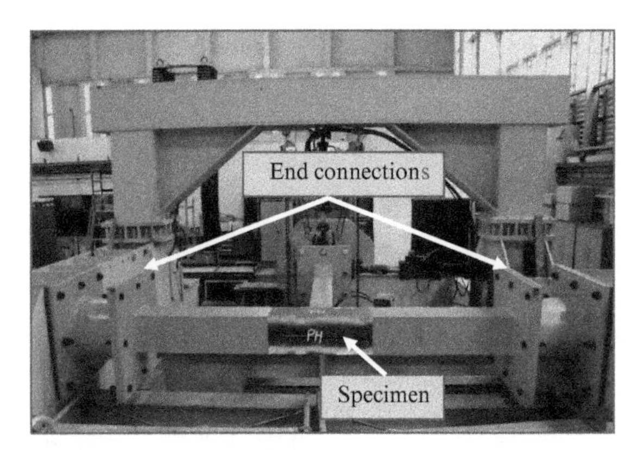

Figure 7.1 Test rig and specimen installation.

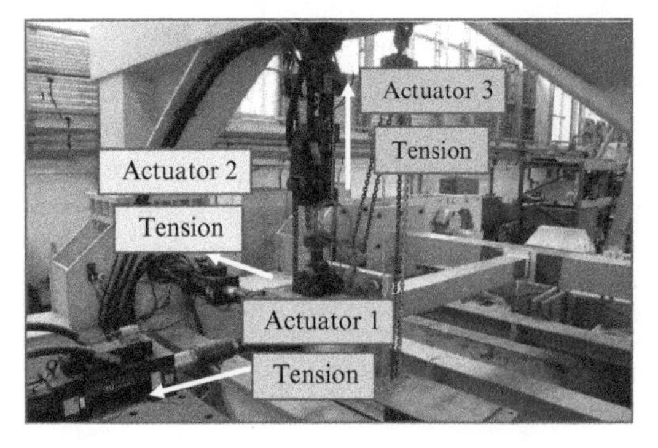

Figure 7.2 Actuators in three directions.

combined load conditions. Figures 7.3 and 7.4 show the dimensions of the joint and end connections.

As mentioned in Chapters 5 and 6, two BHS T-joints were fabricated with HSS RQT701 with a minimum yielding stress of 690 MPa to study the residual stress distribution around the chord weld toe. In this chapter, the preheated T-joint was selected for SCF measurement. Table 7.1 gives geometrical parameters of the T-joints. Considering the fact that the chord length is not long enough to be fixed

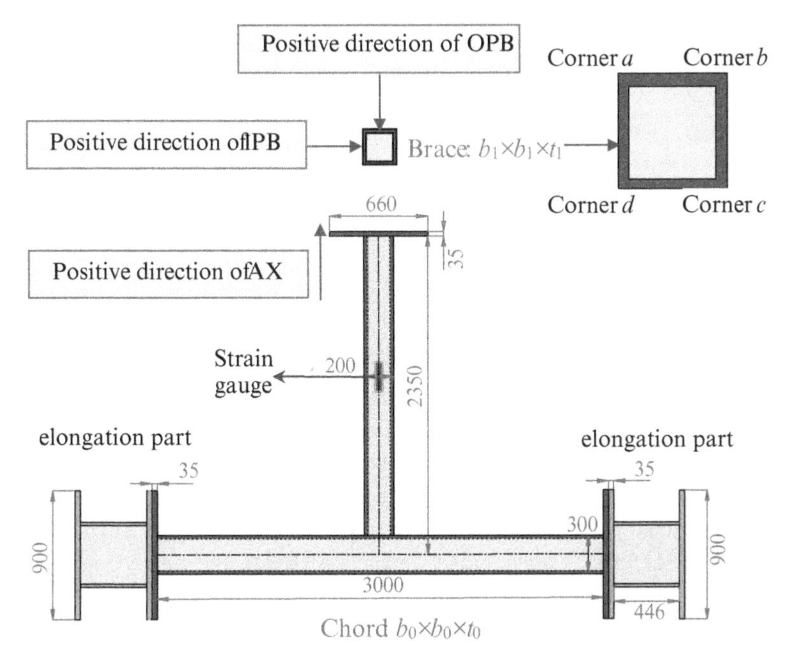

Figure 7.3 The dimensions of the box T-joint and ends connection (all dimension in mm).

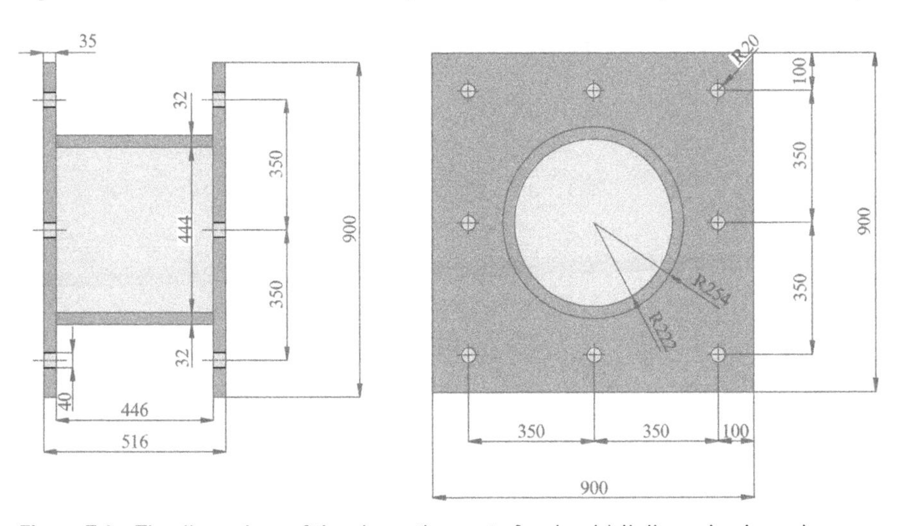

Figure 7.4 The dimensions of the elongation parts for chord (all dimension in mm).

Table 7.1 Geometrical parameters of the T-joints.

Specimen	b_0(mm)	t_0(mm)	b_1(mm)	t_1(mm)	$\beta=\frac{b_1}{b_0}$	$2\gamma=\frac{b_0}{t_0}$	$\tau=\frac{t_1}{t_0}$
Preheated joint	300	12	200	12	0.667	25	1.0

into the test rig, two end supporting sets were fabricated to lengthen the brace. They were designed in such a way that the bending and torsion capacities of the end supporting sets were more than the triple of the bending and torsion capacities of the chord. By doing it this way, the rotation deformation in the connection of the chord and the supporting set can be ignored.

7.3 Strain Gauge Schemes

In order to capture the strain distribution around the brace-to-chord connection, 40 points on the chord (CG1 to CG40) and 28 points on the brace (BG1 to BG28) were selected to capture the strain distribution around the chord-brace intersection (Figure 7.5). On the brace, 28 points are positioned along four brace surface sides and 12 points are located in four corners (Corners *a*, *b*, *c*, and *d*). For every point on the chord, three strain gauges are arrayed along the line perpendicular to the weld toe so that the quadratic extrapolation method can be used to estimate the hot-spot stress at the weld toe. At a selected point on the chord box, the distance from the three strain gauges to the weld toe is 5 mm, 11 mm, and 17 mm, respectively (Figure 7.6).

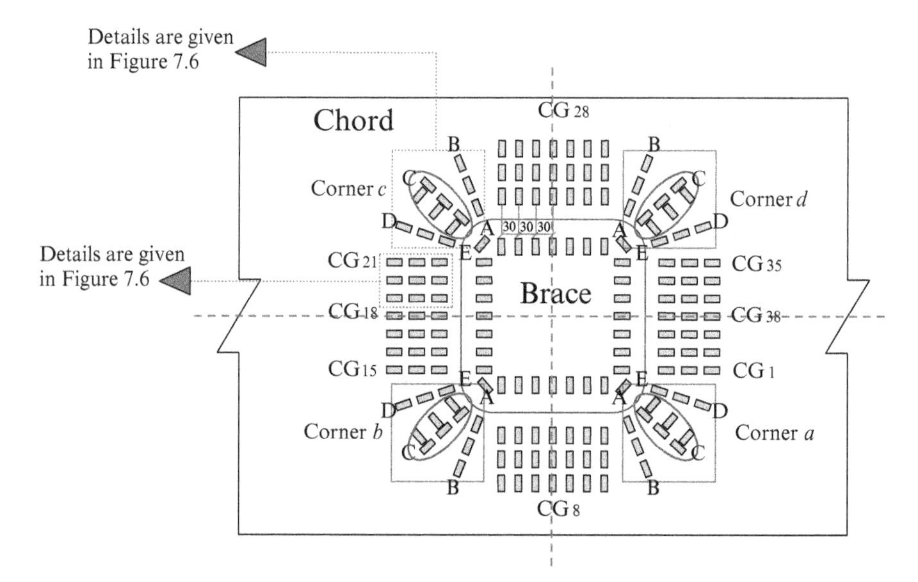

Figure 7.5 Plane view of strain gauge location in preheated joint.

For the all strain gauges that are attached to the specimen, they are chosen to capture the distribution of the perpendicular strain component ($\varepsilon_{\perp}$, the strain perpendicular to the weld toe) around the brace-chord intersection. Figure 7.7 shows the extrapolation zone near the chord weld. At the four corners (Corners *a*, *b*, *c*, and *d*), another four groups of strain gauges that are parallel with the weld toe are glued onto the specimen to capture the strain components parallel to the weld toe (ε_{II}, the strain parallel the weld toe). For all the monitoring points on the chord, the strain $\varepsilon_{\perp}$ at the weld toe is obtained by applying the quadratic extrapolation method based on the readings of the three strain gauges at the point. Note that the strain ε_{II} at the four corners is obtained according to the actual reading in the strain gauges. For the other points, the strain ε_{II} is set to zero.

Finally, another four strain gauges were attached in the center of the four surfaces of the brace at a distance of 1500 mm from the brace weld toe (Figure 7.3). The function of these four strain gauges was to monitor the actual brace load applied. During the test, the strain gauges were connected to five ASW-50 switch boxes and one TDS-530 data logger. The Prolog data collection system was used to record the real-time data (Figures 7.8 and 7.9).

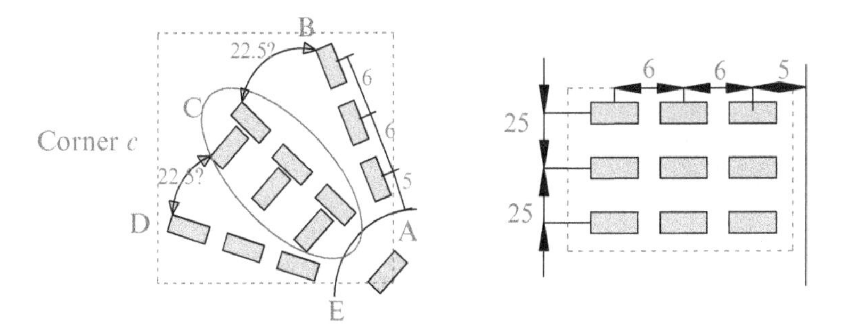

Figure 7.6 Enlarged view of the strain gauge scheme.

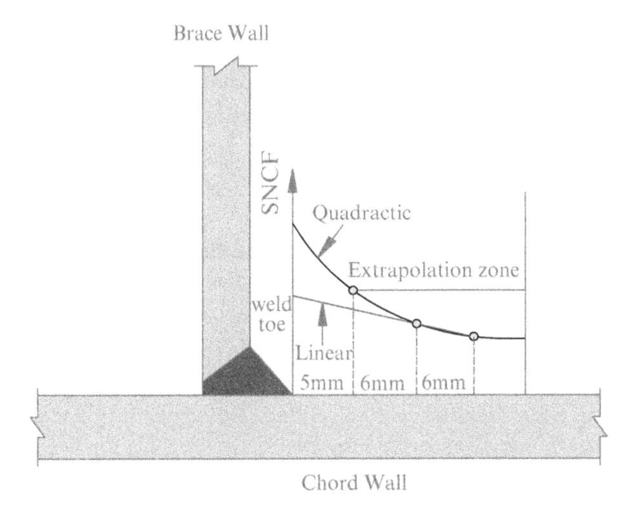

Figure 7.7 Extrapolation zone near the chord weld toe.

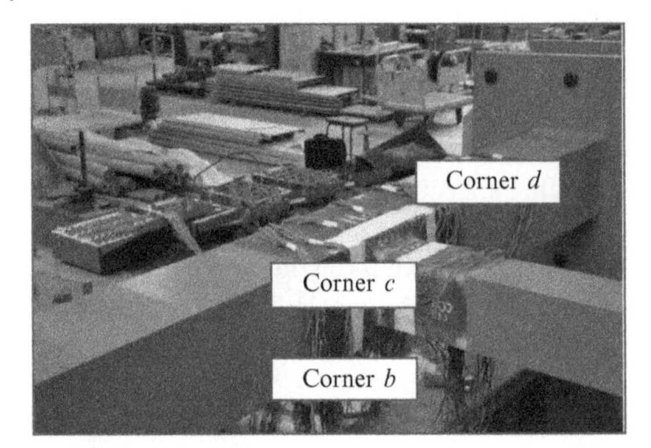

Figure 7.8 Strain gauge put around the intersection.

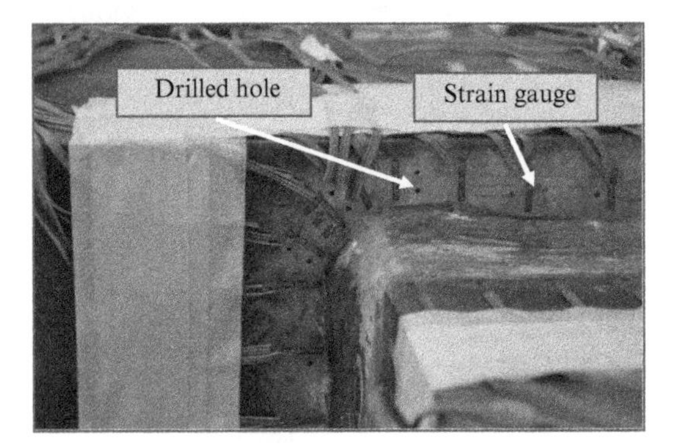

Figure 7.9 A close view of strain gauge put around the intersection.

7.4 Test Procedure

In the static test, a series of load cases consisting of AX, IPB, OPB, and arbitrary combinations of AX, IPB, and OPB were chosen and tested. The basic load cases of AX, IPB, and OPB were employed to obtain the SCFs of the specimens, and to investigate the validity of the published SCF equations.

The first step in the test was to check the stability of the strain gauges. In order to check the stability of the strain gauges, three loading and unloading procedures were carried out to ensure a satisfactory performance of strain gauges and to eliminate the drift or any hysteresis of strain readings. The loads applied for stability-checking were 10KN, 20KN, and 30KN, respectively. After the stability-checking for the strain gauges was carried out, the actual tests were carried out. For each actuator, the load to be applied was pre-programmed and 70% of the material yielding stress was used to calculate the upper limit of the applied load. This ensured that the joint behaved elastically under the maximum load applied. During the static load tests, the actuators were ramped to the maximum predetermined loads in three steps. The actuators were controlled by the means of RS Console software that was developed especially for the servo-hydraulic system. At each load step, the actuators were held in place under active feedback load control while the strain readings are recorded. The loads were then increased to the next load step, and held in place while the strain readings were again taken. After reached the maximum load, the above steps were repeated by releasing the applied load in three steps to zero.

During the test, the strain distribution under the AX load along the weld toe of the brace-chord intersection was investigated first. After that, the strain distribution under the IPB and then the OPB was studied. After the investigation for three basic loads, four combination cases (AX and IPB, AX and OPB, IPB and OPB, AX, IPB and OPB) were explored for the strain distribution near the weld toe.

7.5 Test Results

After the test, strain gauge readings were analyzed and the hot-spot strain (HSSN) was obtained using the quadratic extrapolation method. The quadratic extrapolation was used according to the strain gauge readings along the line perpendicular to the weld toe, with the distance to weld toe being 5 mm, 11 mm, and 17 mm, respectively. Then the strain concentration factor was obtained according to the following equation:

$$SNCF=\frac{HSSN}{SN_{\text{nominal}}} \tag{7.1}$$

where SN_{nominal} is nominal strain computed from the readings of strain gauges at the center of brace sides. After the strain concentration factor was obtained, the SCF could be calculated. According to Hooke's Law for plane stress, there exists,

$$\sigma_{\perp}=\frac{E}{1-\nu^2}(\varepsilon_{\perp}+\varepsilon_{\text{II}}) \tag{7.2}$$

where $\sigma_{\perp}$ is the stress component that is perpendicular to the weld toe; E is Yang's modular; ν is Poisson ratio; and $\varepsilon_{\perp}$, ε_{II} is the strain component perpendicular to the weld toe and parallel to the weld toe, respectively. Since the stress that is perpendicular to the weld toe is adopted in the defining of the SCF, we have

$$SCF=\frac{\sigma_{\perp}}{\sigma_n}=\frac{E}{1-\nu^2}\left(\frac{\varepsilon_{\perp}}{\sigma_n}+\nu\frac{\varepsilon_{\text{II}}}{\sigma_n}\right) \tag{7.3}$$

where σ_{n} is the nominal stress, which can be expressed as $\sigma_{\text{n}} = E\varepsilon_{\text{n}}$ (ε_{n} is nominal strain). Hence,

$$\begin{aligned}SCF&=\frac{1}{1-\nu^2}\left(\frac{\varepsilon_{\perp}}{\varepsilon_n}+\nu\frac{\varepsilon_{\text{II}}}{\varepsilon_n}\right)=\frac{1}{1-\nu^2}\left(SNCF+\nu\frac{\varepsilon_{\text{II}}}{\varepsilon_{\perp}}\frac{\varepsilon_{\perp}}{\varepsilon_n}\right)\\&=\frac{1}{1-\nu^2}\left(SNCF+\nu\frac{\varepsilon_{\text{II}}}{\varepsilon_{\perp}}SNCF\right)=\frac{1}{1-\nu^2}\left(1+\nu\frac{\varepsilon_{\text{II}}}{\varepsilon_{\perp}}\right)SNCF\\&=Snf\cdot SNCF\end{aligned} \tag{7.4}$$

where $Snf = \dfrac{1}{1-\upsilon^2}\left(1+\upsilon\dfrac{\varepsilon_{II}}{\varepsilon_\perp}\right)$, $SNCF = \dfrac{\varepsilon_\perp}{\varepsilon_n}$.

From Eq. 7.5, for the "ideal" situation that $\varepsilon_\perp$ is the only non-zero strain component and $\varepsilon_{II} = 0$, one has $Snf = 1/0.91 \approx 1.10$, which is what was proposed by Dutta. This value has been widely used in RHS joints. According to the parallel strain gauges at the four corners, Snf is approximately equal to 1.2. In this study, the Snf is assumed to be 1.2 at the four corners and 1.1 at the other positions.

The SCF distributions under basic load were then analyzed. The SCF distributions around the chord weld toe for the preheated specimen are shown in Figures 7.10, 7.11 and 7.12. It can be seen that the peak SCF is always located at the corners (*a* to *d*) on the chord under AX, IPB, and OPB.

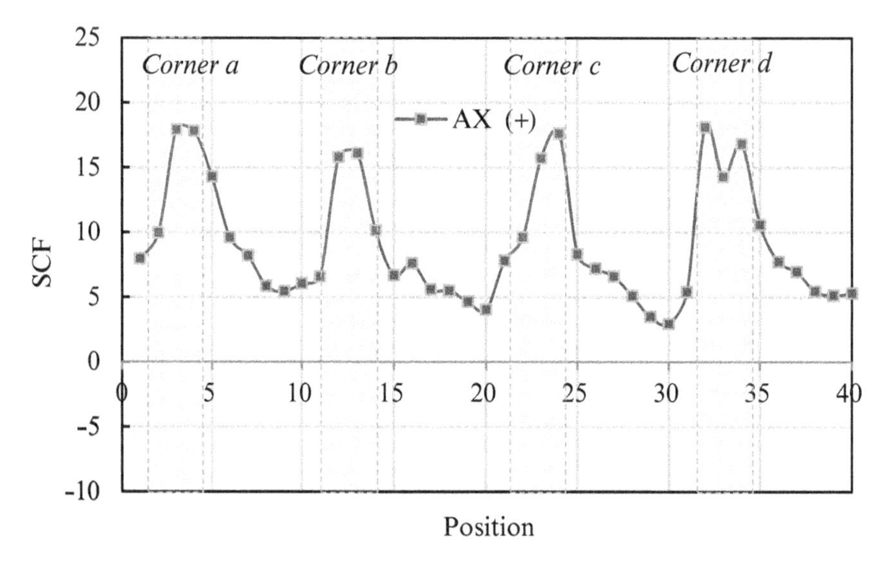

Figure 7.10 SCF distributions on chord box under axial loading.

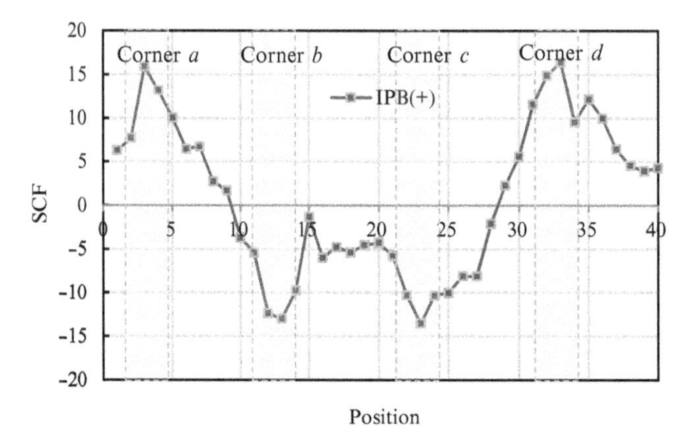

Figure 7.11 SCF distributions on chord box under in-plane bending.

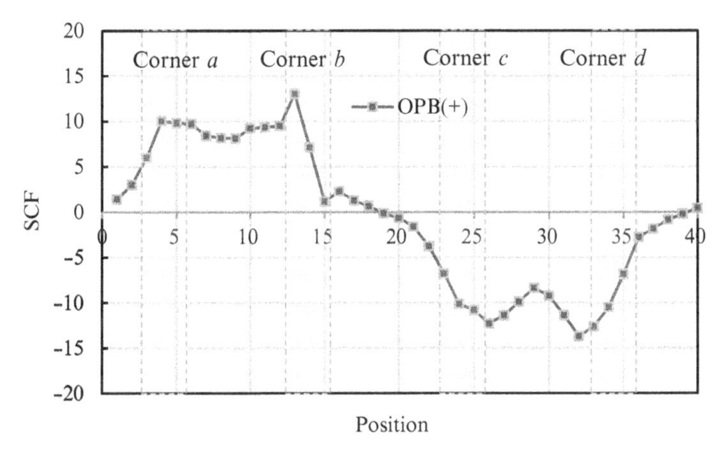

Figure 7.12 SCF distributions on chord box under out-of-plane bending.

7.6 Comparision of Test Results with CIDECT Guide

Five monitoring points (Points A, B, C, D, and E in Figure 7.5) were selected to compare the testing results with CIDECT guide. The difference between the testing result with the CIDECT guide is calculated by Eq. 7.5:

$$Diff = \frac{SCF_{\text{experimental}} - SCF_{CIDECT}}{SCF_{\text{experimental}}} \tag{7.5}$$

Table 7.2 lists the testing results and calculation results by using the equations given by CIDECT guide (Zhao et al. 2000). The SCF values are compared for tensile load and in-plane bending load. Note that in the CIDECT guide, the SCF at Points A and E is provided with the same equation and no formula is given for the out-plane bending load. From Table 7.2, it can be found that for both axial load and in-plane bending, the maximum SCF turns out at B or C. For both loading cases, SCFs at the chord are higher than that at the brace.

Table 7.2 Comparison for SCF values between test results and CIDECT equations.

	AX			IPB		
Location	Exp.	Eqn.	Diff. (%)	Exp.	Eqn.	Diff. (%)
D	9.80	9.61	1.9	9.33	9.38	-0.05
C	16.05	20.94	-30.5	14.70	18.95	-28.9
B	17.30	24.42	-41.2	12.70	14.73	-16.0
A	7.36	12.84	-74.4	6.58	8.42	-28.0
E	9.82	12.84	-30.8	7.17	8.42	-17.4

7.7 Effect of Residual Stress on SCF

The influence of residual stress on SCF is evaluated in this section according to the definition of RSF in Eq. 7.9, which is used to evaluate to the impact of the welding residual stress on the stress concentration effect. Three points (Point D, C, and B) located at the corner of the chord are analyzed. Note that the residual stress at a selected point is the average value based on the residual stress testing results at four corners, which can be found in Chapter 5. It is calculated in such a way because the SCF value in Table 7.2 is the average value based on four corners. In order to make the calculation procedure consistent, both averaged SCF and residual stress values are adopted in analysis.

Figure 7.13 shows the RSF value variations of the ambient temperature specimen under the axial load. When the applied stress is equal to 10 MPa, the *RSF* at Points D, C, and B are equal to 0.71, 0.59, and 0.42, respectively. When the applied stress increases to 50 Mpa, the RSF at Points D, C, and B are equal to 0.33, 0.22, and 0.12, respectively. Therefore, when the applied stress increases from 10 MPa to 50 MPa, the RSF for Points D, C, and B is reduced by 58.5%, 66.7%, and 56.5%, respectively. Thus, the RSF can be reduced greatly when the applied stress increases from 10 MPa to 50 MPa. In other words, the RSF is higher when a lower axial load is applied and the impact of welding residual stress on the stress concentration effect is more important when a lower axial load is applied. When the applied stress increases to 250 MPa, RSF $_{\text{is}}$ reduced to approximately 0.1 for all the points.

Figure 7.14 shows the relationship between RSF and the applied stress for the preheated specimen under the axial load. Similar with Figure 7.13, with increase of axial load, the RSF drops quickly. A noticeable difference between Figures 7.14 and 7.13 is the RSF value for Point B when the axial stress is 10 MPa. When the axial stress is 10 MPa, the RSF for Point B at the ambient temperature joint is 0.42, while the corresponding value at the preheated joint is 0.50. This means that at so low applied stress, the impact of residual stress on the stress concentration is possibly higher at the preheated joint. However, both Figures 7.13 and 7.14 indicate that

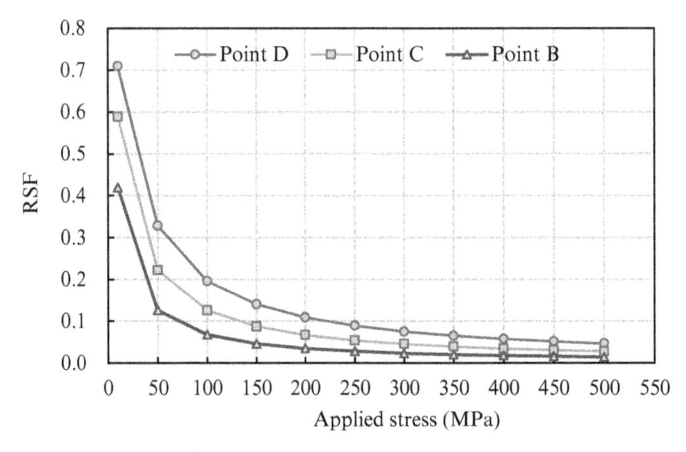

Figure 7.13 RSF distributions at selected points under axial loading (ambient temperature specimen).

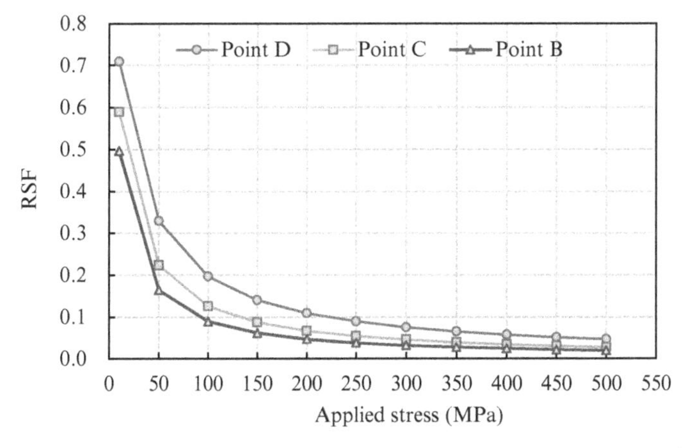

Figure 7.14 RSF distributions at selected points under axial loading (preheated specimen).

when the axial stress is increased, the impact of the residual stress on the SCF is reduced quickly. Therefore, the welding residual stress has more obvious influence on SCF at low applied stress. When the applied stress is higher than 200 MPa (nearly equal to 1/3 of the yielding stress of the material), RSF is less than 0.1, which means the contribution of the welding residual stress on the actual SCF is very small. In this case, the SCF is mainly caused by the joint geometry and loading combination. A similar conclusion can also be drawn from Figures 7.15 and 7.16, which show the RSF distributions at the selected points under in-plane bending.

7.8 Conclusion

In this chapter, a careful experimental study was carried out to investigate the SCF along the joint intersection of preheated box HSS joints. The experimental results show that the maximum SCF should be located in the joint corners. In addition, for

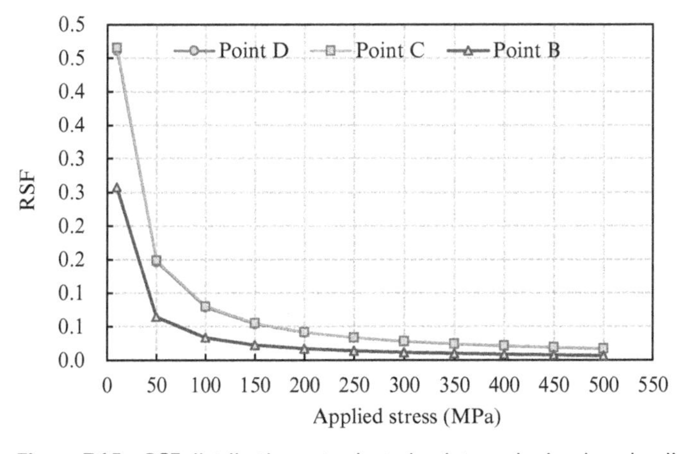

Figure 7.15 RSF distributions at selected points under in-plane loading (ambient temperature specimen).

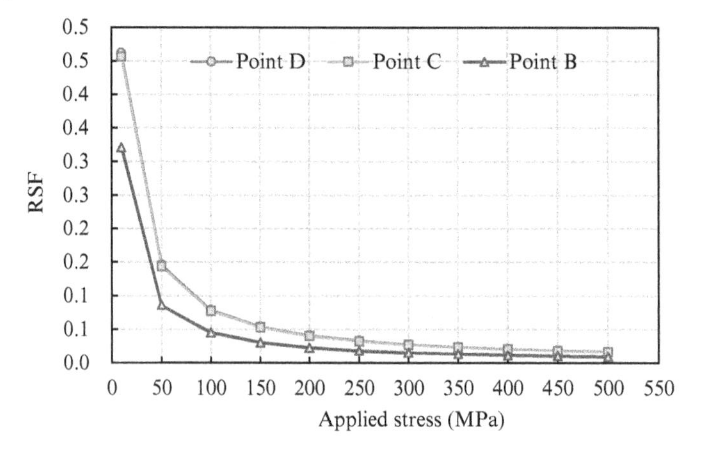

Figure 7.16 RSF distributions at selected points under in-plane bending (preheated specimen).

both axial load and in-plane bending, the maximum SCF turns out at B or C. For both loading cases, SCFs at the chord are higher than that at the brace. By comparing the testing results with the CIDECT, rough agreements can be found at the joint corners for the SCF values.

In this chapter, the impact of the welding residual stress on the stress concentration effect was also studied. The RSF is an acceptable parameter to evaluate the influence of welding residual stress on the stress concentration. It was found that when the axial stress is increased, the impact of the residual stress on the SCF is reduced quickly. The welding residual stress has more obvious influence on SCF at low applied stress. When the applied stress is higher than 200 MPa (nearly equal to 1/3 of the yielding stress of the material), RSF is less than 0.1, which means the contribution of the welding residual stress on the actual SCF is very small.

8

Conclusion and Recommendation

8.1 Introduction

In this study, investigations on the residual stress distributions near the weld toe of HSS plate-to-plate T- and Y-joints and HSS box T-joints were carried out. 18 HSS plate-to-plate T- and Y-joints and 2 HSS box joints were produced and tested.

For the experimental investigation of plate-to-plate T- and Y-joints, two groups of specimens, corresponding to welding preformed at ambient temperature and at a preheating temperature of 100 °C, were fabricated. The effects of preheating and joint geometry on the residual stress distribution near the weld toe were investigated by applying the standard ASTM hole-drilling method for residual stress measurement. Furthermore, a preliminary study was also performed to evaluate the influence of brace plate cutting on the residual stress distribution near the weld toe of the joints.

For the experimental investigation of box T-joints, two specimens with the same geometrical size were fabricated to investigate the welding residual stress distributions of HSS box joints. One specimen was welded at ambient temperature while the other one was preheated to 100 °C before welding.

In addition, a careful sequentially coupled thermo-mechanical modeling procedure was developed for residual stress analysis for the HSS plate-to-plate joints. Both 2D and 3D models were created to investigate the residual stress distribution in the HSS plate-to-plate joints, especially for the residual stress near the weld toe. Validation of the modeling was conducted by comparing with the test results. After validating the accuracy of the modeling procedure, a small-scale parametric study was carried out to investigate the influence of some key welding parameters, such as the boundary condition, the preheating temperature, the number of welding passes, the welding speed, and the welding sequence, on the magnitude and distribution of residual stress. For the HSS box joints, full thermo-mechanical coupled analysis was used to find the residual stress distribution around the joint intersection. In order to reduce the computational cost, only the part (chord length: 0.9m, brace length: 0.3m) near the chord-brace intersection was selected for analysis. Two models corresponding to the specimens (100 °C preheating, welded at ambient temperature) were created and validated by testing results. After that, parametric studies were carried out to

Welded High Strength Steel Structures: Welding Effects and Fatigue Performance, First Edition. Jin Jiang.

investigate the impact of welding parameters, such as preheating temperature, weld speed, and geometrical parameters, such as joint skewed angle and ratio of brace to chord on the residual stress field.

Besides the experimental and numerical work for residual stress, a carefully planned test was organized to analyze SCF values in the HSS plate-to-plate joints. A new factor, the RSF, was proposed to evaluate the effect of residual stress on the stress concentration effect. In addition, a static test on large-scale preheated box T-joints made of HSS RQT701 was conducted to study the SCF distributions under static loading. In the static test, the SCF distributions under three basic loads (AX, IPB, and OPB) were obtained.

8.2 Conclusions

8.2.1 Experimental Studies

18 HSS plate-to-plate joints were tested for the residual stress distribution near the chord plate weld toe. The experimental results indicated that while transverse residual stress with magnitude as high as 1/3 of the yield strength of HSS could appear near the weld toe of the joint, proper preheating could significantly reduce the magnitude of the residual stress. In addition, it was found that the magnitude of residual stress increases as the plate thickness and the intersection angles of the joint increase. Furthermore, the magnitude of the residual stress reduces nonlinearly as the distance from the weld toe increases. By comparing the measurement results obtained from plate-to-plate T-joints with and without brace plate cutting, it was found that mechanical cutting operations near the welding part of a joint could considerably disturb the residual stress distributions along the weld toes. Hence, it is not advisable to perform any cutting prior to residual stress measurement.

For two HSS box T-joints, based on the analysis of the test results for both specimens, several conclusions can be drawn. Firstly, preheating is beneficial to reduce the magnitude of residual stress. Secondly, the residual stress magnitudes along the B1 and B3 sides (parallel to the chord length) are generally smaller than the stresses along the B2 and B4 sides (perpendicular with the chord length). In addition, due to the chord edge effect, the magnitude of residual stress is not always higher for the position closer to the weld toe than that at the further position. The welding residual stress distribution around the intersection of the joints is non-uniform. For both specimens, the residual stress magnitudes at the corners are higher than the values at other positions.

A carefully planned test was organized to analyze SCF values in the HSS plate-to-plate joints and one HSS box joint. In the static tests, both basic loads and combined loads were applied to study the SCF distributions along the chord-brace intersection. The static test results obtained indicate that peak hot-spot stress occurs at the corners of the chord-brace intersection. For both axial load and in-plane bending, the maximum SCF turns out at B or C (the points in the corner, Figure 7.5). For both loading cases, SCFs at the chord are higher than that at the brace. By comparing the testing results with the CIDECT, rough agreements can be found at the joint corners for the SCF values. For the box joints, the impact of the welding residual stress on the stress concentration effect was also studied. It was found that the RSF is an acceptable parameter to evaluate the influence of welding residual stress on the stress concentration. When the axial stress is increased, the impact of the residual stress on the SCF is reduced quickly. The welding residual stress has a more obvious influence on SCF at low applied stress. When the applied stress is higher than 200 MPa (nearly equal to 1/3 of the yielding stress of the material), RSF is less than 0.1, which means the contribution of the welding residual stress on the actual SCF is very small.

8.2.2 Numerical Modeling

2D and 3D procedures of sequentially coupled thermal-stress analysis for a set of HSS plate-to-plate welded joints were developed in the numerical modeling. Element birth and death is an effective way to simulate the addition of filler in the procedure of welding. Sequential coupled thermo-mechanical is an acceptable method for the simulation of residual stress. For the HSS plate-to-plate welded joints, a small-scale parametric study was carried out to evaluate the influence of the mechanical boundary condition, the preheating temperature, the number of welding lumps, the welding speed, and the welding sequence on the residual stress field. The mechanical boundary condition has influence on the residual stress at the weld toe. When both ends in chord plate are fixed, the transverse residual stress at the weld toe is higher than that when pin connections are applied in both ends. When the distance from the weld toe goes further from 15 mm to 50 mm, this influence decreases gradually.

In particular, the preheating treatment can reduce the magnitude of the principal residual stress at the weld toe. This effect is more obvious when the preheating temperature rises to 300 °C. The number of weld lumps is a significant parameter that can affect the residual stress distribution near the weld toe. When more weld lumps are applied for multi-passes welding, a more accurate residual stress at the weld toe can be obtained. Weld speed has a significant influence on the magnitude of residual stress at the weld toe. Higher speed can effectively relieve the residual stress at the weld toe, especially when the weld speed is larger than 3.6 mm/s. Also, when weld filler is added in different sequences, slight changes exist for magnitude of residual stress. The residual stress at the weld toe seems to be lower when the joint angle increases.

For the HSS box joints, 3D procedures of fully coupled thermal-stress analysis were carried out. A modeling procedure for the simulation of welding residual stress for welded box joints was proposed. It was found that the preheating treatment can reduce the magnitude of the principal residual stress at the weld toe. Larger b/c value (the ratio of the brace width to the chord width) is beneficial for reducing the residual stress at the B1 and B3 sides. When b/c is large enough (in other words, the distance between the weld in the chord box and the weld at the chord-brace intersection is close enough, i.e., less than 40 mm), the transverse residual stress between two welds can be reduced due to superposition. However, at a selected point, the residual stress does not show much difference for the cases when b/c is 0.50 and 0.33. The residual stress superposition effect is not obvious when the distance of two welds is large enough.

8.3 Recommendations for Future Research Work

1) Further studies of residual stress measurement by neutron diffraction are highly recommended. In this study, the hole-drilling method was used to measure the welding residual stress near the chord weld toe for both HSS plate-to-plate T/Y-joints and box T-joints. However, due to the limitation of the setup, it is impossible to directly measure the residual stress at the weld toe. In addition, the hole-drilling method is a semi-destructive method, which may bring detrimental influence on the fatigue test. With the neutron diffraction method, the residual stress at the chord weld toe and the brace weld toe could be measured.
2) Fatigue tests for the HSS T-joints and other RHS joint geometry are highly recommended. In the current study, only static tests were carried out for HSS plate-to-plate T/Y-joints and box T-joints. By conducting the fatigue tests, the fatigue performance of the HSS box joints and the impact of the welding residual stress on the fatigue performance could be evaluated.
3) More board numerical investigations for the residual stress at HSS box T-, Y- and K-joints are recommended. In this study, the joint angle, welding speed, b/c value, and preheating temperature parameters were evaluated for their impact on the residual stress. However, the impact of plate thickness was not included in this study. In addition, the modeling for K-joints with different gaps between two braces needs to be investigated in the future.

Summary

In this book, an investigation on the residual stress distributions high strength steel (HSS) welded joints is carried out. Two groups of specimens, corresponding to welding preformed at ambient temperature and at a preheating temperature of 100°C, are fabricated. The effects of preheating and joint geometry on the residual stress distribution near the weld toe are investigated by applying the standard ASTM hole-drilling method for residual stress measurement. A preliminary study is also performed to evaluate the influence of brace plate cutting on the residual stress distribution near the weld toe of the joints.

A carefully sequentially coupled thermal-mechanical modelling procedure is developed for residual stress analysis for the HSS plate-to-plate joints. Both 2D and 3D models were created to investigate the residual stress distribution in the HSS joints. A small-scale parametric study is carried out to investigate the influence of some key welding parameters such as the boundary condition, the preheating temperature, the number of welding pass, the welding speed and the welding sequence on the magnitude and distribution of residual stress. It is found that while transverse residual stress with magnitude as high as one third of the yield strength of HSS could appear near the weld toe of the joint, proper preheating could significantly reduce the magnitude of the residual stress.

Besides the investigation for the residual stress in the HSS plate-to-plate T and Y-joints, the study for welding residual stress in box HSS T-joints is also included. Two box HSS T-joints with the same geometrical size were fabricated to investigate the welding residual stress distributions of HSS box joints. One specimen is welded at ambient temperature while the other one is preheated to 100°C before the welding. It is found that the residual stress at chord weld toe along the surfaces parallel to chord length are generally smaller than along the surfaces perpendicular with chord length.

Fully thermal-mechanical coupled analysis is used in the modeling for the residual stress distribution in box HSS joints. In order to reduce the computational cost, only the chord-brace intersection is selected for analysis. Two models corresponding to the specimens (100°C preheating, welded at ambient temperature) are created and validated by testing results. After that, parametric studies were carried out to investigate the impact of welding parameters such as preheating

Welded High Strength Steel Structures: Welding Effects and Fatigue Performance, First Edition. Jin Jiang.

temperature, weld speeding and geometrical parameters such as joint skewed angle and ratio of brace width to chord width on the residual stress distribution. It is found that the preheating treatment can reduce the magnitude of the principle residual stress at the weld toe. When the ratio of the brace width to the chord width is larger enough, the transverse residual stress between two welds can be reduced due to superposition. In addition, the welding speed has significant influence on the magnitude of residual stress at the weld toe. Higher speed can effectively relieve the residual stress at the weld toe.

Appendix 1 3D Modeling Results of Selected HSS Plate-to-Plate Joints

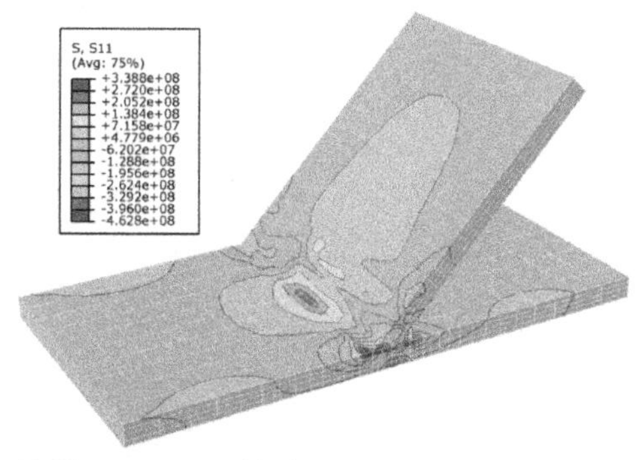

(a). The transverse residual stress at the final state (cooled down)

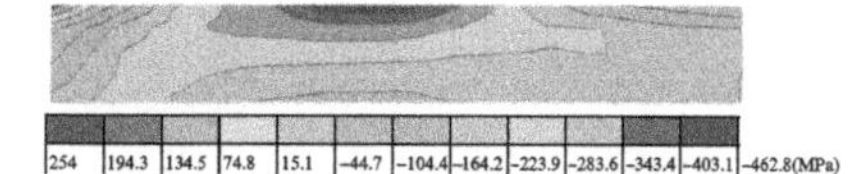

(b). The transverse residual stress at the weld toe (cut by surface)

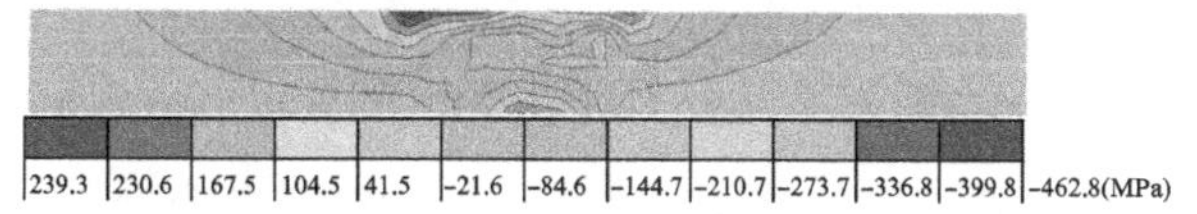

(c). One end cross section of the chord

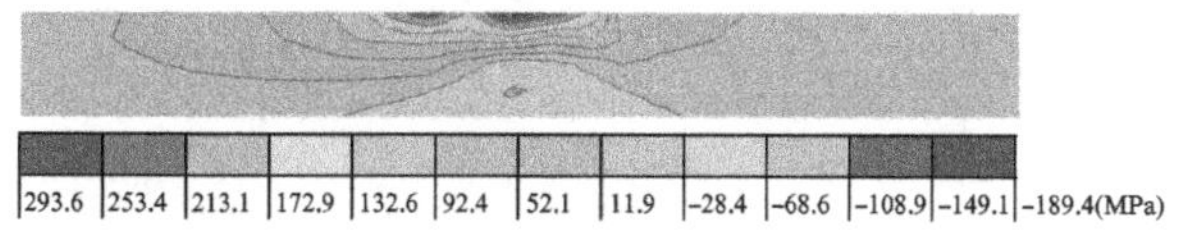

(d). Cross section at middle of chord plate

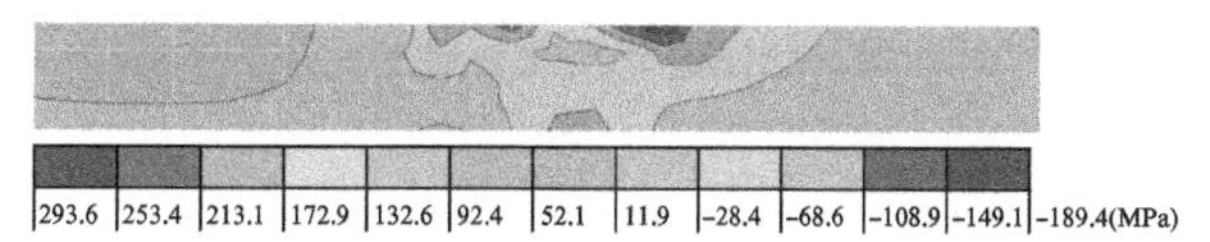

(e). One end cross section of the chord

Figure A1.1 The transverse residual stress profile for different sections at the chord plate ($\theta = 135°$, $t_1 = 12$ mm, ambient temperature, welding direction: case *b*).

Welded High Strength Steel Structures: Welding Effects and Fatigue Performance, First Edition. Jin Jiang.

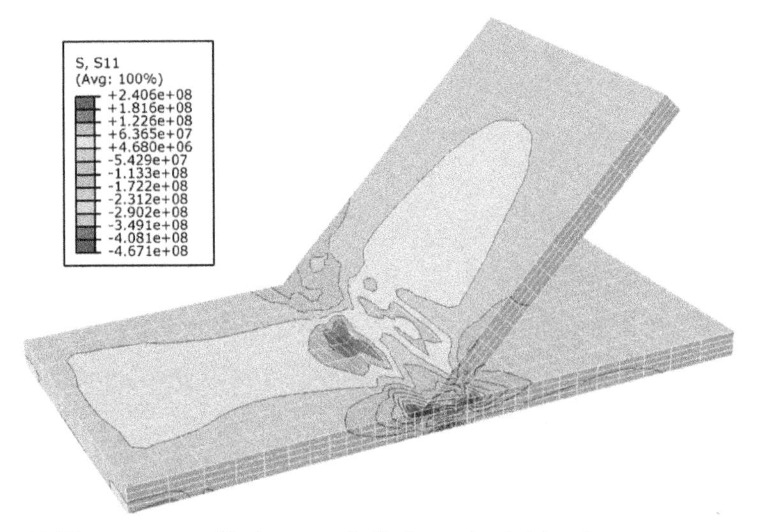

(a). The transverse residual stress at the final state (cooled down)

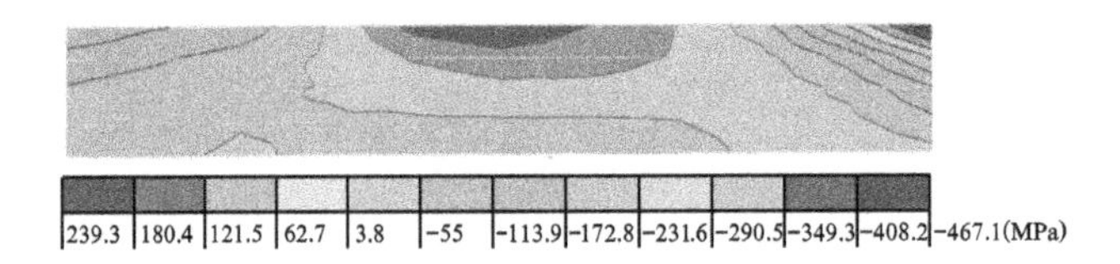

(b). The transverse residual stress at the weld toe (cut by surface)

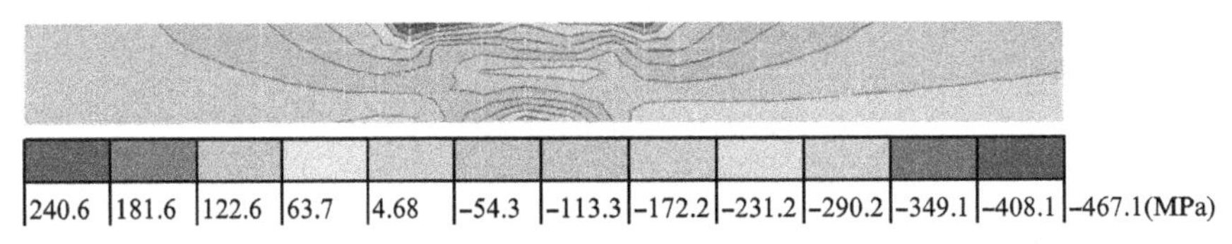

(c). One end cross section of the chord

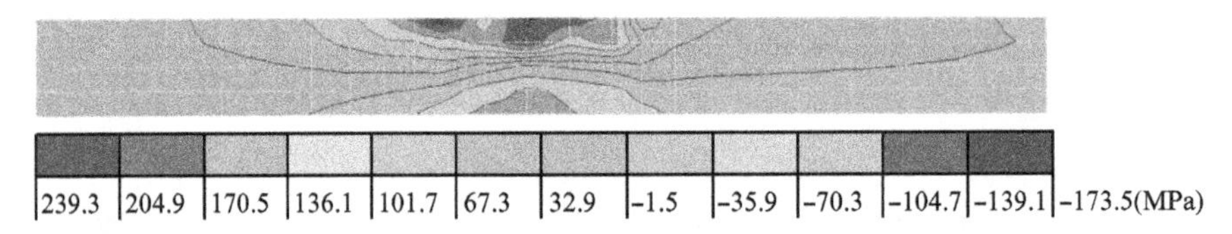

(d). Cross section at middle of chord plate

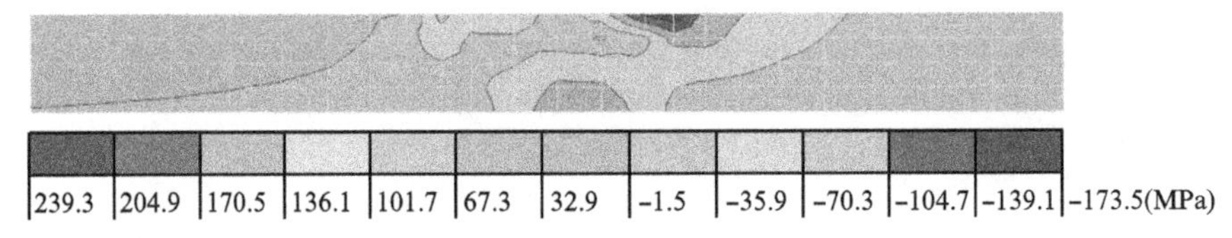

(e). One end cross section of the chord

Figure A1.2 The transverse residual stress profile for different sections at the chord plate (θ = 135°, t_1 = 12 mm, preheated, welding direction: case *b*).

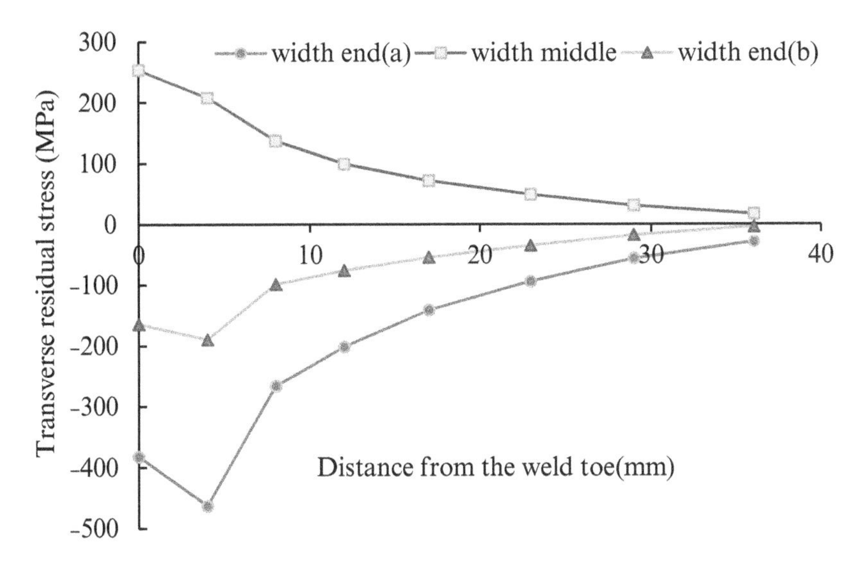

Figure A1.3 Transverse residual stress variation at different locations (Ambient temperature, welding direction: *b*).

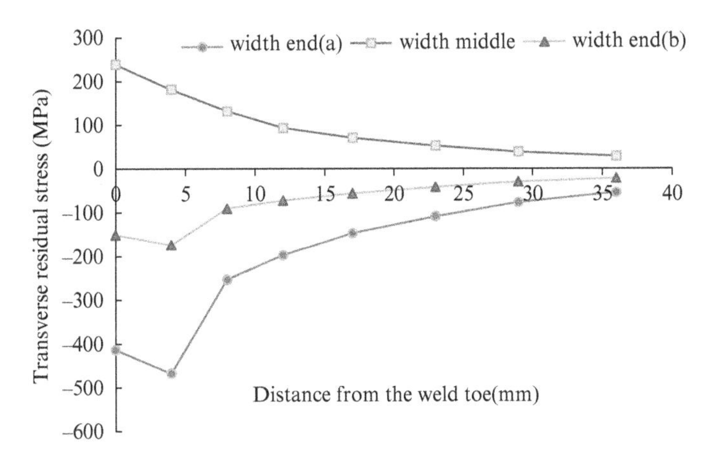

Figure A1.4 Transverse residual stress variation at different locations (Preheated, welding direction: *b*).

Appendix 2 Source Code of Heat Source for 3D HSS Box Hollow Section Joints

```
      SUBROUTINE DFLUX(FLUX,SOL,KSTEP,JINC,TIME,NOEL,NPT,COORDS,
JLTYP,
     1                TEMP,PRESS,SNAME)
CCCCCC THE DOUBLE ELLIPSOIDAL HEAT FLUX MODEL USED (DEVELOPED BY
JIANG JIN) CCCCCCC

      INCLUDE 'ABA_PARAM.INC'
      DIMENSION COORDS(3),FLUX(2),TIME(2)
      CHARACTER*80 SNAME
      REAL*4 u,i,effi,v,q,d
      REAL*4 x0,y0,z0
      REAL*4 a,b,c,aa
      REAL*4 cw,bw,t,w1,w2
CCCCCCCCCCCCCCCCCC THE VARIABLES INCLUDED IN THE SUBROUTINE
CCCCCCCCCCCCCCCCC

C     SOL: Estimated value of the solution variable
C     JSTEP: Step number
C     KINC: Increment number
C     TIME(1): Current value of step time defined only in
transient analysis
C     TIME(2): Current value of total time defined only in
transient analysis
C     NOEL: Element number
C     NPT: Integration point number in the element
C     CORRDS: An array containing the coordinates of this point
C     KLTYP: The flux type used in the model
C     TEMP: Current value of temperature at this integration
point
C     PRESS: Current value of the equivalent pressure stress
C     SNAME: Surface name for a surface-based flux definition
(JLTYP=0).
C     u: The welding voltage used in the joint fabrication
procedure
C     i: The welding current used in the joint fabrication
procedure
```

Welded High Strength Steel Structures: Welding Effects and Fatigue Performance, First Edition. Jin Jiang.

```
C     effi: The welding effiency used in the joint fabrication
procedure
C     v: The welding speed used in the joint fabrication
procedure
C     d: The existing length of the weld filler until the certain
time
C     a,b,c: The shape parameters of the double ellipsoidal model
C     f1,f2: The fractions of the double ellipsoidal model for
the heat
C     cw,bw: The width of the chord and the brace of the T-joint

CCCCCCCCCCCCCCCCCCCCC THE VARIABLES DEFINED BY THE USER
CCCCCCCCCCCCCCCCCCCCC

C    initialization of the parameters used in the model
      u=31
      i=310
      effi=0.8
      v=0.0026
      q=u*i*effi
      d=v*TIME(1)
      cw=0.3
      bw=0.2

C    Define of the start position of the welding
      x=COORDS(1)
      y=COORDS(2)
      z=COORDS(3)
      IF(KSTEP.eq.1) THEN
         x0=0.281
         y0=cw
         z0=0.060
      END IF
      IF(KSTEP.eq.2) THEN
         x0=0.018
         y0=cw
         z0=0.060
      END IF
      IF(KSTEP.eq.3) THEN
         x0=0.281
         y0=0
         z0=0.060
      END IF
      IF(KSTEP.eq.4) THEN
         x0=0.018
         y0=0
         z0=0.060
      END IF
      IF(KSTEP.eq.5) THEN
         x0=0.231
         y0=0.313
         z0=1.6
```

```
      END IF
      IF(KSTEP.eq.6) THEN
         x0=0.068
         y0=0.313
         z0=1.6
      END IF
      IF(KSTEP.eq.7) THEN
         x0=0.213
         y0=0.313
         z0=1.4
      END IF
      IF(KSTEP.eq.8) THEN
         x0=0.068
         y0=0.313
         z0=1.4
      END IF
      IF(KSTEP.eq.9) THEN
         x0=0.231
         y0=0.313
         z0=1.4
      END IF
      IF(KSTEP.eq.10) THEN
         x0=0.25
         y0=0.313
         z0=1.4
      END IF
      IF(KSTEP.eq.11) THEN
         x0=0.25
         y0=0.313
         z0=1.6
      END IF
      IF(KSTEP.eq.12) THEN
         x0=0.05
         y0=0.313
         z0=1.6
      END IF
      IF(KSTEP.eq.13) THEN
         x0=0.05
         y0=0.313
         z0=1.4
      END IF
C     Define of the paramters of the heat flux model
      a=0.016
      b=0.02
      c=0.02
      aa=0.036
      f1=0.7
      PI=3.1415926
      beta=0.0
      vc1=1.06
```

```
      vc1=0.85

            aacos=a/cos(beta/180.0*PI)
       cccos=c*cos(beta/180.0*PI)
       aaacos=aa/cos(beta/180.0*PI)
       cccos=c*cos(beta/180.0*PI)

      heat1=6.0*sqrt(3.0)*q/(a*b*c*PI*sqrt(PI))*f1
      heat2=6.0*sqrt(3.0)*q/(aa*b*c*PI*sqrt(PI))*(2.0-f1)

IF (KSTEP.LE.4) THEN
  shape1=exp(-3.0*(x-x0)**2/aacos**2-3.0*(y-y0)**2/b**2
$      -3.0*(z-z0-d)**2/cccos**2)
  shape2=exp(-3.0*(x-x0)**2/aaacos**2-3.0*(y-y0)**2/b**2
$      -3.0*(z-z0-d)**2/cccos**2)
ELSE IF (KSTEP.ge.5.AND.KSTEP.le.8)THEN
  shape1=exp(-3.0*(x-x0)**2/aacos**2-3.0*(y-y0-d)**2/b**2
$      -3.0*(z-z0)**2/cccos**2)
  shape2=exp(-3.0*(x-x0)**2/aaacos**2-3.0*(y-y0-d)**2/b**2
$      -3.0*(z-z0)**2/cccos**2)
ELSE IF (KSTEP.eq.9)THEN
  shape1=exp(-3.0*(x-x0-d*vc1)**2/aacos**2-3.0*(y-y0)**2/b**2
$      -3.0*(z-z0)**2/cccos**2)
  shape2=exp(-3.0*(x-x0-d*vc1)**2/aaacos**2-3.0*(y-y0)**2/b**2
$      -3.0*(z-z0)**2/cccos**2)
ELSE IF (KSTEP.eq.10)THEN
  shape1=exp(-3.0*(x-x0)**2/aacos**2-3.0*(y-y0)**2/b**2
$      -3.0*(z-z0-d*vc2)**2/cccos**2)
  shape2=exp(-3.0*(x-x0)**2/aaacos**2-3.0*(y-y0)**2/b**2
$      -3.0*(z-z0-d*vc2)**2/cccos**2)
ELSE IF (KSTEP.eq.11)THEN
  shape1=exp(-3.0*(x-x0+d*vc1)**2/aacos**2-3.0*(y-y0)**2/b**2
$      -3.0*(z-z0)**2/cccos**2)
  shape2=exp(-3.0*(x-x0+d*vc1)**2/aaacos**2-3.0*(y-y0)**2/b**2
$      -3.0*(z-z0)**2/cccos**2)
ELSE IF (KSTEP.eq.12)THEN
  shape1=exp(-3.0*(x-x0)**2/aacos**2-3.0*(y-y0)**2/b**2
$      -3.0*(z-z0+d*vc2)**2/cccos**2)
  shape2=exp(-3.0*(x-x0)**2/aaacos**2-3.0*(y-y0)**2/b**2
$      -3.0*(z-z0+d*vc2)**2/cccos**2)
ELSE IF (KSTEP.eq.13)THEN
  shape1=exp(-3.0*(x-x0-d*vc1)**2/aacos**2-3.0*(y-y0)**2/b**2
$      -3.0*(z-z0)**2/cccos**2)
  shape2=exp(-3.0*(x-x0-d*vc1)**2/aaacos**2-3.0*(y-y0)**2/b**2
$      -3.0*(z-z0)**2/cccos**2)

ELSE
END IF

IF (KSTEP.LE.4) THEN
        JLTYP=1
  IF(z .GE.(z0+d)) THEN
```

```
  FLUX(1)=heat1*shape1
  ELSE
  FLUX(1)=heat2*shape2
  ENDIF
ELSE IF (KSTEP.ge.5.AND.KSTEP.le.8)THEN
        JLTYP=1
  IF(y .GE.(y0+d)) THEN
  FLUX(1)=heat1*shape1
  ELSE
  FLUX(1)=heat2*shape2
  ENDIF
ELSE IF (KSTEP.eq.9)THEN
        JLTYP=1
  IF(x .GE.(x0+d)) THEN
  FLUX(1)=heat1*shape1
  ELSE
  FLUX(1)=heat2*shape2
  ENDIF
ELSE IF (KSTEP.eq.10)THEN
        JLTYP=1
  IF(z .GE.(z0+d)) THEN
  FLUX(1)=heat1*shape1
  ELSE
  FLUX(1)=heat2*shape2
  ENDIF
ELSE IF (KSTEP.eq.11)THEN
        JLTYP=1
  IF(x .LE.(x0+d)) THEN
  FLUX(1)=heat1*shape1
  ELSE
  FLUX(1)=heat2*shape2
  ENDIF
ELSE IF (KSTEP.eq.12)THEN
        JLTYP=1
  IF(z .LE.(z0+d)) THEN
  FLUX(1)=heat1*shape1
  ELSE
  FLUX(1)=heat2*shape2
  ENDIF
ELSE IF (KSTEP.eq.13)THEN
        JLTYP=1
  IF(x .GE.(x0+d)) THEN
  FLUX(1)=heat1*shape1
  ELSE
  FLUX(1)=heat2*shape2
  ENDIF
ENDIF
RETURN
END
```

Appendix 3 Source Code of Modeling for 3D HSS Box Hollow Section Joints

```
# The Script Used for Residual Stress Analysis of HSS T/Y Welded
Box Joints #
#         (Developed by Jiang Jin, finally edited at 6, Feb.,
2022)

from abaqus import *
from abaqusConstants import *
import regionToolset

session.viewports['Viewport: 1'].setValues(displayedObject
=None)
user_inputs=getInputs(fields=(('Name of the model:', 'box'),
                    ('Length of the chord:', '0.3'),
                    ('Length of the brace:', '0.2'),
                    ('Thickness of the plate:', '0.012'),
                    ('Skewed angle of the joint','90'),
                    ('Upper weld length of box','0.006'),
                    ('Lower weld length of box','0.013')),
                    label='Please provide the following
information', dialogTitle='Model Parameters')

# If the user left the model name field blank, this needed to
give it a default name.
# This will also be the case if the user hits 'cancel' since
there is no model input.
# If the user enters a character where a number is expected, a
warning is given to the user.

if user_inputs[0]:
   model_name=user_inputs[0]
else:
   print'You did not type in the model name' + \
        'Assuming name-Box joint analysis'
   model_name= 'Box joint analysis'
try:
   box_chord_length=float(user_inputs[1])
except:
```

Welded High Strength Steel Structures: Welding Effect and Fatigue Performance, First Edition. Jin Jiang.

```
    print'You did not type in an integer or float. Assuming a
length of 0.3m'
    box_chord_length=0.3
try:
    box_brace_length=float(user_inputs[2])
except:
    print'You did not type in an integer or float. Assuming a
length of 0.2m'
    box_brace_length=0.2
try:
    box_plate_thickness=float(user_inputs[3])
except:
    print'You did not type in an integer or float. Assuming a
length of 0.012m'
    box_plate_thickness=0.012
try:
    box_skewed_angle=float(user_inputs[4])
except:
    print'You did not type in an integer or float. Assuming an
angle of 90 degrees'
    box_skewed_angle=90
try:
    box_weld_length1=float(user_inputs[5])
except:
    print'You did not type in an integer or float. Assuming a weld
length1 of 0.013m'
    box_weld_length1=0.006
try:
    box_weld_length2=float(user_inputs[6])
except:
    print'You did not type in an integer or float. Assuming a weld
length2 of 0.006m'
    box_weld_length2=0.013

# Input parts defined in the abaqus
# Create the model
mdb.models.changeKey(fromName='Model-1', toName='box')
boxModel=mdb.models['box']
#Create the part
import sketch
import part
t=box_plate_thickness
a=box_skewed_angle
c=box_chord_length
b=box_brace_length
w1=box_weld_length1
w2=box_weld_length2
#(1) Sketch the chord cross section
chordSketch=boxModel.ConstrainedSketch(name='2D Chord Sketch',
sheetSize=10.0)
```

```
xyCoords_1=((0,0),(t,0),(t,c),(0,c),(0,0))
for i in range(len(xyCoords_1)-1):
    chordSketch.Line(point1=xyCoords_1[i],point2=xyCoords_1[i+1])
xyCoords_2=((c-t,0),(c,0),(c,c),(c-t,c),(c-t,0))
for i in range(len(xyCoords_2)-1):
    chordSketch.Line(point1=xyCoords_2[i],point2=xyCoords_2[i+1])
xyCoords_3=((t+w2,0),(c-t-w2,0),(c-t-w1,t),(t+w1,t),(t+
w2,0))
for i in range(len(xyCoords_3)-1):
    chordSketch.Line(point1=xyCoords_3[i],point2=xyCoords_3[i+1])
xyCoords_4=((t+w1,c-t),(c-t-w1,c-t),(c-t-w2,c),(t+w2,
c),(t+w1,c-t))
for i in range(len(xyCoords_4)-1):
    chordSketch.Line(point1=xyCoords_4[i],point2=xyCoords_4[i+1])
chordPart=boxModel.Part(name='Chord', dimensionality=THREE_D,
type=DEFORMABLE_BODY)
chordPart.BaseSolidExtrude(sketch=chordSketch,depth= 3)
#(2) Sketch the chord weld cross section
cweldSketch=boxModel.ConstrainedSketch(name='2D Cweld Sketch',
sheetSize=10.0)
xyCoords_5=((t,0),(t+w2,0),(t+w1,t),(t,t),(t,0))
for i in range(len(xyCoords_1)-1):
    cweldSketch.Line(point1=xyCoords_5[i],point2=xyCoords_5[i+1])
xyCoords_6=((c-t-w2,0),(c-t,0),(c-t,t),(c-t-w1,t),(c-t-w2,0))
for i in range(len(xyCoords_1)-1):
    cweldSketch.Line(point1=xyCoords_6[i],point2=xyCoords_6[i+1])
xyCoords_7=((t,c-t),(t+w1,c-t),(t+w2,c),(t,c),(t,c-t))
for i in range(len(xyCoords_1)-1):
    cweldSketch.Line(point1=xyCoords_7[i],point2=xyCoords_7[i+1])
xyCoords_8=((c-t-w1,c-t),(c-t,c-t),(c-t,c),(c-t-w2,c),(c-t-w1,c-t))
for i in range(len(xyCoords_1)-1):
    cweldSketch.Line(point1=xyCoords_8[i],point2=xyCoords_8[i+1])
cweldPart=boxModel.Part(name='Cweld', dimensionality=THREE_D,
type=DEFORMABLE_BODY)
cweldPart.BaseSolidExtrude(sketch=cweldSketch,depth= 3)
#(3) Sketch the packing of chord cross section
cpackingSketch=boxModel.ConstrainedSketch(name='2D Cpacking
Sketch', sheetSize=10.0)
xyCoords_9=((t,t),(t+0.050,t),(t+0.050,t+0.010),(t,t+0.010),(t,t))
for i in range(len(xyCoords_1)-1):
    cpackingSketch.Line(point1=xyCoords_9[i],point2=xyCoords_9
[i+1])
xyCoords_10=((c-t-0.050,t),(c-t,t),(c-t,t+0.010),(c-t-0.050,-
t+0.010),(c-t-0.050,t))
for i in range(len(xyCoords_1)-1):
    cpackingSketch.Line(point1=xyCoords_10[i],point2=xyCoords_10
[i+1])
xyCoords_11=((t,c-t-0.010),(t+0.050,c-t-0.010),(t+0.050,c-t),(t,
c-t),(t,c-t-0.010))
```

```
for i in range(len(xyCoords_1)-1):
    cpackingSketch.Line(point1=xyCoords_11[i],point2=xyCoords_11
[i+1])
xyCoords_12=((c-t-0.050,c-t-0.010),(c-t,c-t-0.010),(c-t,c-t),(c-
t-0.050,c-t),(c-t-0.050,c-t-0.010))
for i in range(len(xyCoords_1)-1):
    cpackingSketch.Line(point1=xyCoords_12[i],point2=xyCoords_12
[i+1])
cpackingPart=boxModel.Part(name='Cpacking', dimensionality=THREE_D,
type=DEFORMABLE_BODY)
cpackingPart.BaseSolidExtrude(sketch=cpackingSketch,depth= 3)

#(4) Sketch the brace cross section
braceSketch=boxModel.ConstrainedSketch(name='2D Brace Sketch',
sheetSize=10.0)
xyCoords_b1=((0,0),(t,0),(t,b),(0,b),(0,0))
for i in range(len(xyCoords_b1)-1):
    braceSketch.Line(point1=xyCoords_b1[i],point2=xyCoords_
b1[i+1])
xyCoords_b2=((b-t,0),(b,0),(b,b),(b-t,b),(b-t,0))
for i in range(len(xyCoords_b2)-1):
    braceSketch.Line(point1=xyCoords_b2[i],point2=xyCoords_
b2[i+1])
xyCoords_b3=((t+w2,0),(b-t-w2,0),(b-t-w1,t),(t+w1,t),(t+w2,0))
for i in range(len(xyCoords_b3)-1):
    braceSketch.Line(point1=xyCoords_b3[i],point2=xyCoords_
b3[i+1])
xyCoords_b4=((t+w1,b-t),(b-t-w1,b-t),(b-t-w2,b),(t+w2,b),
(t+w1,b-t))
for i in range(len(xyCoords_b4)-1):
    braceSketch.Line(point1=xyCoords_b4[i],point2=xyCoords_
b4[i+1])
bracePart=boxModel.Part(name='Brace', dimensionality=THREE_D,
type=DEFORMABLE_BODY)
bracePart.BaseSolidExtrude(sketch=braceSketch,depth= 2)

#(5) Sketch the brace weld cross section
bweldSketch=boxModel.ConstrainedSketch(name='2D Bweld Sketch',
sheetSize=10.0)
xyCoords_b5=((t,0),(t+w2,0),(t+w1,t),(t,t),(t,0))
for i in range(len(xyCoords_b5)-1):
    bweldSketch.Line(point1=xyCoords_b5[i],point2=xyCoords_
b5[i+1])
xyCoords_b6=((b-t-w2,0),(b-t,0),(b-t,t),(b-t-w1,t),(b-t-w2,0))
for i in range(len(xyCoords_b6)-1):
    bweldSketch.Line(point1=xyCoords_b6[i],point2=xyCoords_
b6[i+1])
xyCoords_b7=((t,b-t),(t+w1,b-t),(t+w2,b),(t,b),(t,b-t))
for i in range(len(xyCoords_b7)-1):
```

```
    bweldSketch.Line(point1=xyCoords_b7[i],point2=xyCoords_
b7[i+1])
xyCoords_b8=((b-t-w1,b-t),(b-t,b-t),(b-t,b),(b-t-w2,b),(b-t-w1,b-
t))
for i in range(len(xyCoords_b8)-1):
    bweldSketch.Line(point1=xyCoords_b8[i],point2=xyCoords_
b8[i+1])
bweldPart=boxModel.Part(name='Bweld', dimensionality=THREE_D,
type=DEFORMABLE_BODY)
bweldPart.BaseSolidExtrude(sketch=bweldSketch,depth= 2)
#(6) Sketch the packing of brace cross section
bpackingSketch=boxModel.ConstrainedSketch(name='2D Bpacking
Sketch', sheetSize=10.0)
xyCoords_b9=((t,t),(t+0.050,t),(t+0.050,t+0.010),(t,t+0.010),(t
,t))
for i in range(len(xyCoords_b9)-1):
    bpackingSketch.Line(point1=xyCoords_b9[i],point2=xyCoords_
b9[i+1])
xyCoords_b10=((b-t-0.050,t),(b-t,t),(b-t,t+0.010),(b-t-0.050,t-
+0.010),(b-t-0.050,t))
for i in range(len(xyCoords_b10)-1):
    bpackingSketch.Line(point1=xyCoords_b10[i],point2=xyCoords_
b10[i+1])
xyCoords_b11=((t,b-t-0.010),(t+0.050,b-t-0.010),(t+0.050,b-t),(t,-
b-t),(t,b-t-0.010))
for i in range(len(xyCoords_b11)-1):
    bpackingSketch.Line(point1=xyCoords_b11[i],point2=xyCoords_
b11[i+1])
xyCoords_b12=((b-t-0.050,b-t-0.010),(b-t,b-t-0.010),(b-t,b-t),(b-
t-0.050,b-t),(b-t-0.050,b-t-0.010))
for i in range(len(xyCoords_b12)-1):
    bpackingSketch.Line(point1=xyCoords_b12[i],point2=xyCoords_
b12[i+1])
bpackingPart=boxModel.Part(name='Bpacking', dimensionality=THREE
_D, type=DEFORMABLE_BODY)
bpackingPart.BaseSolidExtrude(sketch=bpackingSketch,depth= 2)

#(7) Sketch the welding for the intersection of the box joint
weldingSketch=boxModel.ConstrainedSketch(name='2D Welding
Sketch', sheetSize=10.0)
xyCoords_w1=((0,0),(0.010+t,0),(0.010+t,0.006+t),(0.010,0.006
+t),(0,0))
for i in range(len(xyCoords_w1)-1):
    weldingSketch.Line(point1=xyCoords_w1[i],point2=xyCoords_
w1[i+1])
xyCoords_w2=((c-(c-b)/2-t,0),(c-(c-b)/2+0.010,0),(c-(c-b)/-
2,0.006+t),(c-(c-b)/2-t,0.006+t),(c-(c-b)/2-t,0))
for i in range(len(xyCoords_w2)-1):
    weldingSketch.Line(point1=xyCoords_w2[i],point2=xyCoords_
w2[i+1])
```

```
weldingPart=boxModel.Part(name='Welding', dimensionality=
THREE_D, type=DEFORMABLE_BODY)
weldingPart.BaseSolidExtrude(sketch=weldingSketch,depth= b-2*t)
#Create the material properties for the thermo-mechanical coupled
analysis
#Mass density, Young's modulus, thermal conductivity, elastic,
and plastic properties will be defined
import material
boxMaterial=boxModel.Material(name='RQT701')
boxMaterial.Density(temperatureDependency=ON,table=(
(7.850E-9,30),(7.835E-9,100),(7.801E-9,200),
(7.764E-9,300),(7.726E-9,400),(7.687E-9,500),
(7.648E-9,600),(7.600E-9,700),(7.551E-9,800),
(7.501E-9,900),(7.448E-9,1000),(7.393E-9,1100),(7.336E-9,1200)))

boxMaterial.Conductivity(temperatureDependency=ON, table=(
(53.17,30),(50.67,100),(47.34,200),(44.01,300),
(40.68,400),(37.35,500),(34.02,600),(30.69,700),
(27.36,800),(27.30,900),(27.30,1000),(27.30,1100),(27.30,1200)))

boxMaterial.Expansion(temperatureDependency=ON, table=(
(3.68E-4,30),(9.98E-4,100),(2.3E-3,200),(3.7E-3,300),
(5.2E-3,400),(6.8E-3,500),(8.4E-3,600),(1.01E-2,700),
(1.1E-2,800),(1.18E-2,900),(1.38E-2,1000),(1.58E-
2,1100),(1.78E-2,1200)))

boxMaterial.Elastic(temperatureDependency=ON, table=(
(2.1E5,0.300,30),(2.09E5,0.305,100),(1.97E5,0.311,200),
(1.9E5,0.313,300),(1.81E5,0.315,400),(1.7E5,0.318,500),
(1.6E5,0.324,600),(1.48E5,0.336,700),(1.42E5,0.34,800),
(1.42E5,0.340,900),(1.42E5,0.34,1000),(1.42E5,0.34,1100),(1.4
2E5,0.34,1200)))

boxMaterial.Plastic(temperatureDependency=ON, table=(
(730,0,30),(840,0.1,30),(702,0,100),(814,0.1,100),(654,0,200),
(768,0.1,200),(586,0,300),(681,0.1,300),(499,0,400),(577,0,400),
(393,0,500),(458,0.1,500),(270,0,600),(323,0.1,600),(145,0,700),
(191,0.1,700),(135,0,800),(167,0.1,800),(95.5,0,900),(119,0.1,900),
(23.3,0,1000),(41.8,0.1,1000),(23.3,0,1200),(41.8,0.1,1200)))

boxMaterial.SpecificHeat(temperatureDependency=ON, table=(
(4.433E8,30),(4.876E8,100),(5.300E8,200),(5.647E8,300),
(6.059E8,400),(6.665E8,500),(7.602E8,600),(1.0E8,700),
(1.4E9,750),(8.033E8,800),(6.5E8,900),(6.5E8,1000),(6.5E8,1100),
(6.5E8,1200)))

#Create a section and assign the box joint to it
import section
boxSection=boxModel.HomogeneousSolidSection(name='Box Section',
material='RQT701')
chord_region=(chordPart.cells,)
```

```
chordPart.SectionAssignment(region=chord_region, sectionName='Box
Section')
cweld_region=(cweldPart.cells,)
cweldPart.SectionAssignment(region=cweld_region, sectionName='Box
Section')
cpacking_region=(cpackingPart.cells,)
cpackingPart.SectionAssignment(region=cpacking_region,
sectionName='Box Section')
brace_region=(bracePart.cells,)
bracePart.SectionAssignment(region=brace_region, sectionName='Box
Section')
bweld_region=(bweldPart.cells,)
bweldPart.SectionAssignment(region=bweld_region, sectionName='Box
Section')
bpacking_region=(bpackingPart.cells,)
bpackingPart.SectionAssignment(region=bpacking_region,
sectionName='Box Section')
welding_region=(weldingPart.cells,)
weldingPart.SectionAssignment(region=welding_region,
sectionName='Box Section')

#Create the assembly for whole parts
import assembly
boxAssembly=boxModel.rootAssembly
chordInstance=boxAssembly.Instance(name='chord instance',
part=chordPart,dependent=OFF)
cweldInstance=boxAssembly.Instance(name='cweld instance',
part=cweldPart,dependent=OFF)
cpackingInstance=boxAssembly.Instance(name='cpacking instance',pa
rt=cpackingPart,dependent=OFF)
braceInstance=boxAssembly.Instance(name='brace instance',
part=bracePart,dependent=OFF)
bweldInstance=boxAssembly.Instance(name='bweld instance',
part=bweldPart,dependent=OFF)
bpackingInstance=boxAssembly.Instance(name='bpacking instance',pa
rt=bpackingPart,dependent=OFF)
weldingInstance=boxAssembly.Instance(name='welding instance',part=w
eldingPart,dependent=OFF)

# Boolean operation for the created parts
boxAssembly.InstanceFromBooleanMerge(name='mychord', instances=(
        chordInstance, cweldInstance,cpackingInstance),keepInters
ections=ON,
        originalInstances=SUPPRESS, domain=GEOMETRY)
boxAssembly. InstanceFromBooleanMerge(name='mybrace', instances=(
        braceInstance, bweldInstance, bpackingInstance,),
keepIntersections=ON,
        originalInstances=SUPPRESS, domain=GEOMETRY)

boxAssembly.makeIndependent(instances=(boxAssembly.instances['my
chord-1'], ))
```

```
boxAssembly.makeIndependent(instances=(boxAssembly.instances['my
brace-1'], ))
boxAssembly.features.changeKey(fromName='mybrace-1',toName='mybr
ace')
boxAssembly.features.changeKey(fromName='mychord-1',toName='mych
ord')
boxAssembly.deleteFeatures(('chord instance', 'cweld instance',
'cpacking instance',
                            'brace instance', 'bweld instance',
'bpacking instance', ))
mybraceInstance=boxAssembly.instances['mybrace']
mychordInstance=boxAssembly.instances['mychord']

# Adjust the relevant positioning for instances
boxAssembly.translate(instanceList=('mybrace', ),vector=
(0.0,c,0.0))
boxAssembly.rotate(instanceList=('mybrace', ), axisPoint=(0.0, c,
0.0),
   axisDirection=(b, 0.0, 0.0), angle=-90.0)
boxAssembly.translate(instanceList=('mybrace', ), vector=(c-b,
0.0, 1.5+b/2))
boxAssembly.translate(instanceList=('mybrace', ), vector=(-abs(b-
c)/2, 0.0, 0))
session.viewports['Viewport: 1'].setValues(displayedObject=boxAs
sembly)

#Create the fully coupled thermo-mechanical analysis steps
import step
boxModel.CoupledTempDisplacementStep(name='Step-1', previous
='Initial',
        timePeriod=1153.8, maxNumInc=10000, initialInc=10.0,
minInc=1e-09,
         maxInc=50.0, deltmx=50.0, nlgeom=ON)
boxModel.CoupledTempDisplacementStep(name='Step-2',
previous='Step-1',
        timePeriod=1153.8, maxNumInc=10000, initialInc=10.0,
minInc=1e-09,
        maxInc=50.0, deltmx=50.0, nlgeom=ON)
boxModel.CoupledTempDisplacementStep(name='Step-3',
previous='Step-2',
        timePeriod=1153.8, maxNumInc=10000, initialInc=10.0,
minInc=1e-09,
        maxInc=50.0, deltmx=50.0, nlgeom=ON)
boxModel.CoupledTempDisplacementStep(name='Step-4',
previous='Step-3',
        timePeriod=1153.8, maxNumInc=10000, initialInc=10.0,
minInc=1e-09,
        maxInc=50.0, deltmx=50.0, nlgeom=ON)
boxModel.CoupledTempDisplacementStep(name='Step-5',
previous='Step-4',
```

```
        timePeriod=769.2, maxNumInc=10000, initialInc=10.0,
minInc=1e-09,
        maxInc=50.0, deltmx=50.0, nlgeom=ON)
boxModel.CoupledTempDisplacementStep(name='Step-6',
previous='Step-5',
        timePeriod=769.2, maxNumInc=10000, initialInc=10.0,
minInc=1e-09,
        maxInc=50.0, deltmx=50.0, nlgeom=ON)
boxModel.CoupledTempDisplacementStep(name='Step-7',
previous='Step-6',
        timePeriod=769.2, maxNumInc=10000, initialInc=10.0,
minInc=1e-09,
        maxInc=50.0, deltmx=50.0, nlgeom=ON)
boxModel.CoupledTempDisplacementStep(name='Step-8',
previous='Step-7',
        timePeriod=769.2, maxNumInc=10000, initialInc=10.0,
minInc=1e-09,
        maxInc=50.0, deltmx=50.0, nlgeom=ON)
boxModel.CoupledTempDisplacementStep(name='Step-9',
previous='Step-8',
        timePeriod=50, maxNumInc=10000, initialInc=5.0,
minInc=1e-09,
        maxInc=10.0, deltmx=50.0, nlgeom=ON)
boxModel.CoupledTempDisplacementStep(name='Step-10',
previous='Step-9',
        timePeriod=50, maxNumInc=10000, initialInc=5.0,
minInc=1e-09,
        maxInc=10.0, deltmx=50.0, nlgeom=ON)
boxModel.CoupledTempDisplacementStep(name='Step-11', previous='S
tep-10',
        timePeriod=50, maxNumInc=10000, initialInc=5.0,
minInc=1e-09,
        maxInc=10.0, deltmx=50.0, nlgeom=ON)
boxModel.CoupledTempDisplacementStep(name='Step-12', previous='S
tep-11',
        timePeriod=50, maxNumInc=10000, initialInc=5.0,
minInc=1e-09,
        maxInc=10.0, deltmx=50.0, nlgeom=ON)
boxModel.fieldOutputRequests.changeKey(fromName='F-Output-1',toN
ame='field')
boxModel.FieldOutputRequest(name='fieldoutput',createStepName='S
tep-1',
        variables=('S', 'E', 'NT', 'TEMP', 'HFL'),frequency=20)
#Apply the boundary conditions for the objective model
end_face_1_pt_1=(c/2,t/2,0)
end_face_1_pt_2=(c/2,c-t/2,0)
end_face_1_pt_3=(t/2,c/2,0)
end_face_1_pt_4=(c-t/2,c/2,0)
end_face_1_pt_5=(t+w2/2,t/2,0)
```

```
end_face_1_pt_6=(c-t-w2/2,t/2,0)
end_face_1_pt_7=(t+w2/2,c-t/2,0)
end_face_1_pt_8=(c-t-w2/2,c-t/2,0)
end_face_1_pt_9=(t+0.025,t+0.005,0)
end_face_1_pt_10=(c-t-0.025,t+0.005,0)
end_face_1_pt_11=(t+0.025,c-t-0.005,0)
end_face_1_pt_12=(c-t-0.025,c-t-0.005,0)
end_face_1=mychordInstance.faces.findAt(((end_face_1_pt_1),),
                                        ((end_face_1_pt_2),),
                                        ((end_face_1_pt_3),),
                                        ((end_face_1_pt_4),),
                                        ((end_face_1_pt_5),),
                                        ((end_face_1_pt_6),),
                                        ((end_face_1_pt_7),),
                                        ((end_face_1_pt_8),),
                                        ((end_face_1_pt_9),),
                                        ((end_face_1_pt_10),),
                                        ((end_face_1_pt_11),),
                                        ((end_face_1_pt_12),))
regionbd_1 = regionToolset.Region(faces=end_face_1)
boxModel.PinnedBC(name='BC-1', createStepName='Step-1',
region=regionbd_1)
end_face_2_pt_1=(c/2,t/2,3)
end_face_2_pt_2=(c/2,c-t/2,3)
end_face_2_pt_3=(t/2,c/2,3)
end_face_2_pt_4=(c-t/2,c/2,3)
end_face_2_pt_5=(t+w2/2,t/2,3)
end_face_2_pt_6=(c-t-w2/2,t/2,3)
end_face_2_pt_7=(t+w2/2,c-t/2,3)
end_face_2_pt_8=(c-t-w2/2,c-t/2,3)
end_face_2_pt_9=(t+0.025,t+0.005,3)
end_face_2_pt_10=(c-t-0.025,t+0.005,3)
end_face_2_pt_11=(t+0.025,c-t-0.005,3)
end_face_2_pt_12=(c-t-0.025,c-t-0.005,3)
end_face_2=mychordInstance.faces.findAt(((end_face_2_pt_1),),
                                        ((end_face_2_pt_2),),
                                        ((end_face_2_pt_3),),
                                        ((end_face_2_pt_4),),
                                        ((end_face_2_pt_5),),
                                        ((end_face_2_pt_6),),
                                        ((end_face_2_pt_7),),
                                        ((end_face_2_pt_8),),
                                        ((end_face_2_pt_9),),
                                        ((end_face_2_pt_10),),
                                        ((end_face_2_pt_11),),
                                        ((end_face_2_pt_12),))
regionbd_2 = regionToolset.Region(faces=end_face_2)
boxModel.PinnedBC(name='BC-2', createStepName='Step-1',
region=regionbd_2)
```

```
box_joint_region=((mychordInstance.cells,),(mybraceInstance.cel
ls,),)
boxModel.Temperature(name='Predefined Field-1',createStepName=
'Initial',
         region=box_joint_region, distributionType=UNIFORM,
         crossSectionDistribution=CONSTANT_THROUGH_THICKNESS,
magnitudes=(30.0,))

#Create interactions (convection and radiation)
convection_face_1_pt_1=(0,c/2,1.5)
convection_face_1_pt_2=(t,c/2,1.5)
convection_face_1_pt_3=(c-t,c/2,1.5)
convection_face_1_pt_4=(c,c/2,1.5)

convection_face_1_pt_5=(c/2,0,1.5)
convection_face_1_pt_6=(c/2,t,1.5)
convection_face_1_pt_7=(c/2,c-t,1.5)
convection_face_1_pt_8=(c/2,c,1.5)

convection_face_1_pt_9=(t+0.025,t+0.010,1.5)
convection_face_1_pt_10=(c-t-0.025,t+0.010,1.5)
convection_face_1_pt_11=(t+0.025,c-t-0.010,1.5)
convection_face_1_pt_12=(c-t-0.025,c-t-0.010,1.5)
convection_face_1_pt_13=(t+0.050,t+0.005,1.5)
convection_face_1_pt_14=(c-t-0.050,t+0.005,1.5)
convection_face_1_pt_15=(t+0.050,c-t-0.005,1.5)
convection_face_1_pt_16=(c-t-0.050,c-t-0.005,1.5)

convection_face_1_pt_17=(t+w2/2,0,1.5)
convection_face_1_pt_18=(c-t-w2/2,0,1.5)
convection_face_1_pt_19=(t+w2/2,c,1.5)
convection_face_1_pt_20=(c-t-w2/2,c,1.5)

convection_face_1_pt_21=(t/2,0,1.5)
convection_face_1_pt_22=(c-t/2,0,1.5)
convection_face_1_pt_23=(t/2,c,1.5)
convection_face_1_pt_24=(c-t/2,c,1.5)
convection_face_1=mychordInstance.faces.findAt(
                                   ((convection_face_1_pt_1),),
                                   ((convection_face_1_pt_2),),
                                   ((convection_face_1_pt_3),),
                                   ((convection_face_1_pt_4),),
                                   ((convection_face_1_pt_5),),
                                   ((convection_face_1_pt_6),),
                                   ((convection_face_1_pt_7),),
                                   ((convection_face_1_pt_8),),
                                   ((convection_face_1_pt_9),),
                                   ((convection_face_1_pt_10),),
                                   ((convection_face_1_pt_11),),
                                   ((convection_face_1_pt_12),),
                                   ((convection_face_1_pt_13),),
```

```
                                                  ((convection_face_1_pt_14),),
                                                  ((convection_face_1_pt_15),),
                                                  ((convection_face_1_pt_16),),
                                                  ((convection_face_1_pt_17),),
                                                  ((convection_face_1_pt_18),),
                                                  ((convection_face_1_pt_19),),
                                                  ((convection_face_1_pt_20),),
                                                  ((convection_face_1_pt_21),),
                                                  ((convection_face_1_pt_22),),
                                                  ((convection_face_1_pt_23),),
                                                  ((convection_face_1_pt_24),))

convection_face_2_pt_1=(abs(c-b)/2+0,c+1.0,1.5)
convection_face_2_pt_2=(abs(c-b)/2+t,c+1.0,1.5)
convection_face_2_pt_3=(abs(c-b)/2+b-t,c+1.0,1.5)
convection_face_2_pt_4=(abs(c-b)/2+b,c+1.0,1.5)

convection_face_2_pt_5=(abs(c-b)/2+b/2,c+1.0,1.5-b/2)
convection_face_2_pt_6=(abs(c-b)/2+b/2,c+1.0,1.5-b/2+t)
convection_face_2_pt_7=(abs(c-b)/2+b/2,c+1.0,1.5+b/2-t)
convection_face_2_pt_8=(abs(c-b)/2+b/2,c+1.0,1.5+b/2)

convection_face_2_pt_9=(abs(c-b)/2+t/2,c+1.0,1.5-b/2)
convection_face_2_pt_10=(abs(c-b)/2+t+w2/2,c+1.0,1.5-b/2)
convection_face_2_pt_11=(abs(c-b)/2+b-t/2,c+1.0,1.5-b/2)
convection_face_2_pt_12=(abs(c-b)/2+b-t-w2/2,c+1.0,1.5-b/2)

convection_face_2_pt_13=(abs(c-b)/2+t/2,c+1.0,1.5+b/2)
convection_face_2_pt_14=(abs(c-b)/2+t+w2/2,c+1.0,1.5+b/2)
convection_face_2_pt_15=(abs(c-b)/2+b-t/2,c+1.0,1.5+b/2)
convection_face_2_pt_16=(abs(c-b)/2+b-t-w2/2,c+1.0,1.5+b/2)

convection_face_2_pt_17=(abs(c-b)/2+t+0.025,c+1.0,1.5-b/2
+t+0.010)
convection_face_2_pt_18=(abs(c-b)/2+b-t-0.025,c+1.0,1.5-b/2
+t+0.010)
convection_face_2_pt_19=(abs(c-b)/2+t+0.025,c+1.0,1.5+b/2-
t-0.010)
convection_face_2_pt_20=(abs(c-b)/2+b-t-0.025,c+1.0,1.5+b/2-
t-0.010)
convection_face_2_pt_21=(abs(c-b)/2+t+0.050,c+1.0,1.5-b/2
+t+0.005)
convection_face_2_pt_22=(abs(c-b)/2+b-t-0.050,c+1.0,1.5-b/2
+t+0.005)
convection_face_2_pt_23=(abs(c-b)/2+t+0.050,c+1.0,1.5+b/2-
t-0.005)
convection_face_2_pt_24=(abs(c-b)/2+b-t-0.050,c+1.0,1.5+b/2-
t-0.005)
convection_face_2=mybraceInstance.faces.findAt(
                                                 ((convection_face_2_pt_1),),
                                                 ((convection_face_2_pt_2),),
```

```
                                          ((convection_face_2_pt_3),),
                                          ((convection_face_2_pt_4),),
                                          ((convection_face_2_pt_5),),
                                          ((convection_face_2_pt_6),),
                                          ((convection_face_2_pt_7),),
                                          ((convection_face_2_pt_8),),
                                          ((convection_face_2_pt_9),),
                                          ((convection_face_2_pt_10),),
                                          ((convection_face_2_pt_11),),
                                          ((convection_face_2_pt_12),),
                                          ((convection_face_2_pt_13),),
                                          ((convection_face_2_pt_14),),
                                          ((convection_face_2_pt_15),),
                                          ((convection_face_2_pt_16),),
                                          ((convection_face_2_pt_17),),
                                          ((convection_face_2_pt_18),),
                                          ((convection_face_2_pt_19),),
                                          ((convection_face_2_pt_20),),
                                          ((convection_face_2_pt_21),),
                                          ((convection_face_2_pt_22),),
                                          ((convection_face_2_pt_23),),
                                          ((convection_face_2_pt_24),))

convection_surface_region_1 =
regionToolset.Region(side1Faces=convection_face_1)
convection_surface_region_2 =
regionToolset.Region(side1Faces=convection_face_2)
boxModel.FilmCondition(name='convection-1', createStepName='St
ep-1',
          surface=convection_surface_region_1,
definition=EMBEDDED_COEFF, filmCoeff=0.012,
          filmCoeffAmplitude='', sinkTemperature=30.0, sinkAmpli-
tude='')
boxModel.FilmCondition(name='convection-2', createStepName='St
ep-5',
          surface=convection_surface_region_2,
definition=EMBEDDED_COEFF, filmCoeff=0.012,
          filmCoeffAmplitude='', sinkTemperature=30.0, sinkAmpli-
tude='')
boxModel.RadiationToAmbient(name='radiation-1', createStepName='
Step-1',
         surface=convection_surface_region_1, radiationType=AMBIE
NT,
         distributionType=UNIFORM, field='', emissivity=0.02, am-
bientTemperature=30.0,
         ambientTemperatureAmp='')
boxModel.RadiationToAmbient(name='radiation-2', createStepName='
Step-5',
```

```
        surface=convection_surface_region_2, radiationType=AMBIE
NT,
        distributionType=UNIFORM, field='', emissivity=0.02, am-
bientTemperature=30.0,
        ambientTemperatureAmp='')
boxModel.setValues(absoluteZero=-273.15, stefanBoltzmann=5.67E-8)

#Meshing for the joint assembly
import mesh
patition_face_1=boxAssembly.DatumPlaneByTwoPoint(point1=(c/2
,c,1.5-b/2), point2=(c/2,c,1.5-b/2-0.04))
patition_face_2=boxAssembly.DatumPlaneByTwoPoint(point1=(c/2
,c,1.5-b/2-0.04), point2=(c/2,c,1.5-b/2-0.44))
patition_face_3=boxAssembly.DatumPlaneByTwoPoint(point1=(c/2,c,1.
5+b/2), point2=(c/2,c,1.5+b/2+0.04))
patition_face_4=boxAssembly.DatumPlaneByTwoPoint(point1=(c/2,c,1.
5+b/2+0.04), point2=(c/2,c,1.5+b/2+0.44))
boxAssembly.PartitionCellByDatumPlane(datumPlane=patition_face_1,
cells=mychordInstance.cells)
```

References

ABAQUS (2009). 6.9 Theory manual. *Dassault Systèmes Simulia Corp*. Providence, US.

Acevedo, C. (2009). Residual stress estimation of welded tubular K-joints under fatigue loads. *Proceeding of the 12th International Conference on Fracture (ICF12)*. Ottawa, Canda.

Acevedo, C. (2011). Influence of residual stresses on fatigue response of welded tubular joints. *École polytechnique fédérale de Lausanne (EPFL).*

Acevedo, C. and Nussbaumer, A. (2010). Influence of welding residual stresses on stable crack growth in tubular K-joints under compressive fatigue loadings. *13th International Symposium on Tubular Structures*, 557–565. Hong Kong.

Agerskov, H. (2000). Fatigue in steel structures under random loading. *Journal of Construction Steel Research* 53 (3): 283–305.

Agerskov, H. and Petersen, R.I. (1998). An investigation on fatigue in high-strength steel offshore structures. *Welding in the World* 41 (4): 328–342.

Anami, K. and Miki, C. (2001). Fatigue strength of welded joints made of high strength steels. *Progress in Structural Engineering and Materials* 2001 (3): 86–94.

Anderson, L.F. (2000). Residual stress and deformations in steel structures. *Department of Navel Architecture and Offshore Engineering, Technical University of Denmark.*

API (1993). *A.P.I.,* Recommended practice for planning, designing and constructing fixed offshore platforms. *American Petroleum Institute*. Washington DC, US.

AS/NZS (2004). Structural steel welding, in Part 4: welding of high strength quenched and tempered steel. *Standards Australia*.

ASTM (2005). Standard Specification for High Yield Strength, Quenched and Tempered Alloy Steel Plate, Suitable for Welding. ASTM International. West Conshohocken, United States.

ASTM (2008). Standard test method for determining residual stresses by hole-drilling strain-gage method (E837-08). ASTM International. West Conshohocken, US.

AWS (2006). Specification for low-alloy steel electrodes for shielded metal arc welding. American Welding Society. Miami, US.

AWS (2008). Structural welding code-steel *(AWS D1.1)*. American Welding Society. Miami, US.

Beaney, E.M. and Procter, E. (1974). A critical evaluation of the center hole technique for the measyrement of residual stresses. *Strain* 10 (1): 7–14.

Billingham, J., Sharp, J.V., Spurrier, J., and Kilgallon, P.J. (2003). Review of the performance of high strength steels used offshore. *Research Report 105*. Cranfield University for the Health and Safety Executive.

Billingham, J. and Spurrier, J. (1995). The influence of welding on the performance of high strength steel offshore.

Bjorhovde, R. (2004). Development and use of high performance steel. *Journal of Construction Steel Research* 60 (2004): 393–400.

Bozidar, L. (2010). Quenching Theory and Technology. Hoboken: CRC Press.

Brickstad, B. and Josefson, B.L. (1998). A parametric study of residual stresses in multi-pass butt-welded stainless steel pipes. *International Journal of Pressure Vessels and Piping*, 75 (1): 11–25.

Brule, A. and Kirstein, O. (2006). Residual stress diffractometer KOWARI at the Australian research reactor OPAL: status of the project. *Physica B* 385–386 (2006): 1040–1042.

Brust, F.W. and Rybicki, E.F. (1981). A Computational Model of backlay welding for controlling residual stresses in welded pipes. *Journal of Pressure Vessel Technology* 103 (3): 226–232.

BSI (1992). Tensile testing of metallic materials Part 5: method of test at elevated temperatures. British Standard Institute (BSI). London, UK.

BSI (2011). Welding recommendations for welding of metallic materials. *In Part 2, Arc Welding of Ferritic Steels*. London, UK.

Chen, H. (2005). Fatigue resistance of high performance steel (HPS) details. University of Alberta.

Chen, H., Grondin, G.Y., and Drive, R.G. (2003). Fatigue properties of high performance steel. *1st International Conference on Fatigue Damage of Materials*, 181–191. Toronto, Canada.

Claphama, L., Abdullah, K., Jeswiet, J.J. et al. (2004). Neutron diffraction residual stress mapping in same gauge and differential gauge tailor-welded blanks. *Journal of Materials Processing Technology* 148 (2004): 177–185.

Clarin, M. (2004). High strength steel local buckling and residual stresses. Department of Civil and Environmental Engineering, Lulea University of Technology.

Den (1990). Offshore installations: guidance on design and construction. UK: Depart of Energy.

DNV (2008a). Fatigue Design of Offshore Steel Structures *(DNV-RP-C203)*. DET NORSKE VERITAS.

DNV (2008b). Offshore Standard of Metallic Materials(DNV-OS-B101). DET NORSKE VERITAS.

NORSOK STANDARD (2004). Design of steel structures(NORSOK standard N-004), Norway.

Easterling, K.E. (1992). Introduction to the Physical Metallurgy of Welding. Butterworth-Heinemann.

EC3 (1993a). Design of steel structure-Part 1.1: general rules and rules for building. European Committee for Standardisation (CEN).

EC3 (1993b). Design of steel structure-Part 1.9: fatigue. European Committee for Standardisation (CEN).

EC3 (2005). Design of steel structures Part 1-2: general rules — structural fire design. European Committee for Standardisation (CEN).

Etube, L.S. (2001). Fatigue and fracture mechanics of offshore structures. *Engineering Research Series*. Professional Engineering Publishing Limited. London and Bury St Edmunds, UK.

Free, A., Porter Goff, R.F.D. (1989). Predicting residual stresses in multi-pass weldments with the finite element method. *Computers and Structures* 32: 365–378.

Fitzpatrick, M.E., Fry, A.T., Holdway, P. et al. (2005). Determination of residual stresses by X-ary Diffraction (Issus2). *Measurement Good Practice Guide (No.52)*. National Physical Laboratory. Teddington, Middlesex, UK.

Flaman, M.T. (1982). Investigation of ultra-high speed drilling for residual stress measurements by center hole method. *Experimental Mechanics* 22 (1): 26–30.

Goldak, J., Chakravarti, A. and Bibby, M. (1984). A new finite element model for welding heat sources. *Metallurgical transactions B* 15: 299–305.

Goldak, J.A. and Akhlaghi M. (2005). Computational welding mechanics. Springer Science & Business Media.

Granjon, H. (1991). *Fundamentals of Welding Metallurgy*. Woodhead Publishing.

Grant, P.V., Lord, J.D., and Whitehead, P.S. (2002). The measurement of residual stresses by the incremental hole drilling technique. *Measurement Good Practice Guide (No.53)*. National Physical Laboratory. Teddington, Middlesex, UK.

Hagiwara, Y., Kadono, A., Suzuki, T. et al. (1995). Application of HT780 high strength steel plate to structural member of super high-rise building: part 2 Reliability inspection of the structure. *Proceedings of the 5th East Asia-Pacific Conference on Structural Engineering and Construction*, 2289–2294. Gold Coast.

Hajro, I., Pasic, O., and Burzic, Z. (2010). Characterization of welded joints on high-strength structural steels S690QL and S890QL. *2nd South-East European IIW International Congress*, Sofia.

Healy, J. and Billingham, J. (1995). Metallurgical considerations of the high yield to ultimate ratio in high strength steels for use in offshore engineering. *14th International Conference of OMAE*.

Hong, J.K., Tsai, C.L., and Dong, P. (1998). Assessment of numerical procedures for residual stress analysis of multipass welds. *Welding Journal* 77 (9): 372–382.

HSE (1999). Static strength of cracked high strength steel tubular joints. Health and Safety Executive.

IACS (1999). Fatigue assessment of ship structures. International Association of Classification Societies LTD.

IIW (1999). Recommended fatigue design procedure for hollow section joints: part 1 Hot spot stress method for nodal joints. Lisbon, Portugal.

ISO (2008). Petroleum and natural gas industries—Fix steel offshore structures.

Jang, G.C., Chang, K.H., and Lee, C.H. (2007). Characteristics of the residual stress distribution in welded tubular T-joints. *Journal of Mechanical Science and Technology* 21 (10): 1714–1719.

Kandil, F.A., Lord, J.D., Fry, A.T., and Grant, P.V. (2001). A review of residual stress measurement methods: guideto technique selection in NPL report MATC(A)O4. National Physical Laboratory. Teddington, Middlesex, UK.

Kaufmann, H., Sonsino, C.M., and Demofonti, G. (2007). High-strength steels in welded state for light-weight constructions under high and variable stress peaks. *Steel Research International* 79 (5): 382–389.

King, R.N. (1998). A review of fatigue crack growth rates in air and seawater. Offshore Technology Report.

Kou, S. and Le, Y. (1988). Welding parameters and the grain structure of weld metal -A thermodynamic consideration. Metallurgical Transactions A 19: 1075–1082.

Krebs, J. and Kassner, M. (2007). Influence of Welding Residual Stresses on Fatigue Design of Welded Joints and Components. Weld World 51: 54–68.

Lancaster, J. (1997). *Handbook of Structural Welding*. Milton Keynes, England: Abington Publishing.

Law, M., Kirstein, O., and Luzin, V. (2010). An assessment of the effect of cutting welded samples on residual stress measurements by chill modelling. *Journal of Strain Analysis* 45 (8): 567–573.

Lee, H.Y., Biglari, F.R., Wimpory, R., and Nikbin, K.M. (2006). Treatment of residual stress in failure assessment of procedure. *Engineering Fracture Mechanics* 73 (13): 1755–1771.

Lindgren, L.E. (2001a). Finite element modelling and simulation of welding Part 1: increased complexity. *Journal of Thermal Stresses* 24: 141–192.

Lindgren, L.E. (2001b). Finite element modelling and simulation of welding Part 2: improved material modelling. *Journal of Thermal Stresses* 24: 195–231.

Lindgren, L.E. and Karlsson, L. (1988). Deformations and stresses in welding of shell structures. *International Journal for Numerical Methods in Engineering* 25: 635–655.

Lu, J. (1996). *Handbook of Measurement of Residual Stresses*. The Fairmont Press.

Maddox, S.J. (1991). *Fatigue Strength of Welded Structures*, 2e. Abington Hall, England: Abington Publishing, Woodhead Publishing Ltd.

Masubuchi, K. (1980). Analysis of welded structures: residual stresses, distortion, and their consequences. In: *International Series on Materials Science and Technology*. Pergamon Press.

Mather, J. (1934). Determination of initial stresses by measuring the deformation around drilled holed. *Transactions ASME* 56 (4): 349–254.

Michaleris, P. (1996). Residual stress distributions for multi-pass welds in pressure vessel and piping components. *ASME Pressure Vessel and Piping Conference: Residual Stress in Design, Fabrication, Assessment and Repair*. 327.

Micro-Measurements (2005). Non-destructive testing—Standard test method for determining residual stresses by neutron diffraction.

Micro-Measurements (2007). Measurement of residual stresses by the hole-drilling strain gage method, Tech Note TN-503, 19–33.

Miki, C., Homma, K., and Tominaga, T. (2002). High strength and high performance steels and their use in bridge structures. *Journal of Construction Steel Research* 58 (2002): 3–20.

Mochuzihi, H., Yamashita, T., and Fukasawa, T. (1995). Application of HT780 high strength steel plate to structural member of super high-rise building: part 1 Development of high strength steel with heavy gauge and welding process. Proceedings of the 5th East Asia-Pacific Conference on Structural Engineering and Construction, 2283–2288. Gold Coast.

Norwegian standard (2004). Design of steel structures (N-004).

Oettel, R. (2000). The determination of uncertainties in residual stress measurement. *Manual of Code of Practice for the Determination of Uncertainties in Mechanical Tests on Metallic Materials (No.15)*, Dresden, Germany.

Pang, H.L. (1989). Residual stress measurement in a cruciform welded joint using hole drilling and strain gauges. *Strain* 1989: 7–14.

Park, M.J., Yang, H.N., Jang, D.Y. et al. (2004). Residual stress measurement on welded specimen by neutron diffraction. *Journal of Materials Processing Technology* 155–156 (2004): 1171–1177.

Payne, J.G. and Porter-Goff, R.F.D. (1986). Experimental residual stress distributions in welded tubular T-nodes. *The Institution of Mechanical Engineers (I Mech E)* C134/86: 109–116.

Pavelic, V., Tanbakuchi, R., Uyehara, O.A., and Myers, P.S. (1969). *Welding Journal Research Supplement* 48: 295s–305s.

Pedersen, N.T. and Agerskov, H. (1991). Fatigue life prediction of offshore steel structures under stochastic loading, in Series *R*. Department of Structural Engineering, Technical University of Denmark.

Petersen, R.I., Agerskov, H., and Lopez, M.L. (1996). Fatigue life of high strength steel offshore tubular joints. Technical University of Denmark.

Prevéy, P.S. (1996). Current application of X-ray diffraction residual stress measurement. Lambda Technologies. 1–8.

Radaj, D. and Zhang, S. (1992). Stress intensity factors for spot welds between plates of dissimilar materials. Engineering Fracture Mechanics 42 (3): 407–426.

Ravi, S., Balasubramanian, V., Babu, S., and Nasser, S.N. (2004). Assessment of some factors influencing the fatigue life of strength mis-matched HSLA steel weldments. *Materials and Design* 25 (3): 125–135.

Rendler, N.J. and Vigness, I. (1966). Hole-drilling strain-gage method of measuring residual stresses. *Experimental Mechanics* 6 (12): 577–586.

Rosenthal, D. (1946). The theory of moving sources of heat and its application to metal treatments. Transactions of the American Society of Mechanical Engineers 68 (8): 849–865.

Prior, F and Maurer, K L. (1995). Fatigue strength properties of welded joints of high-strength weldable structural steel; Schwingfestigkeitseigenschaften von Schweissverbindungen hochfester, schweissgeeigneter Baustaehle. Germany: N. Web.

Rybicki, E.F., Schmueser, D.W., Stonesifer, R.W., and Groom, J.J. (1978). A finite-element model for residual stresses and deflections in girth-butt welded pipes. *Journal of Pressure Vessel Technology* 100 (3): 256–262.

Rybicki, E.F. and Stonesifer, R.B. (1979). Computation of residual stresses due to multipass welds in piping systems. *Journal of Pressure Vessel Technology* 101 (2): 149–154.

Schajer, G.S. (1988a). Measurement of non-uniform residual stresses using the hole-drilling method. Part 1: stress calculation procedures. *Journal of Engineering Materials and Technology* 110 (4): 338–349.

Schajer, G.S. (1988b). Measurement of non-uniform residual stresses using the hole-drilling method. Part 2: practical application of the integral method. *Journal of Engineering Materials and Technology* 110 (4): 344–349.

Schajer, G.S. (2009). Hole-drilling residual stress measurements at 75: origins, advances, opportunities. *Experimental Mechanics* 50 (2): 245–253.

Sedlacek, G. and Muller, C. (2005). The use of very high strength steels in metallic construction. *Proceedings of 1st International Conference Super-high Strength Steels.*

Sharp, J.V., Billingham, J., and Stacey, A. (1999). Performance of high strength steels used in jack-ups. *Marine Structures* 12: 349–370.

Shim, Y., Feng, Z., Lee, S. et al. (1992). Determination of residual stresses in thick-section weldments. *Welding Journal* 71 (9): 305–312.

Sonsino, C.M. (2009). Effect of residual stresses on the fatigue behaviour of welded joints depending on loading conditions and weld geometry. *International Journal of Fatigue* 31 (1): 88–101.

Sonsino, C.M., Lagoda, T., and Demofonti, G. (2004). Damage accumulation under variable amplitude loading of welded medium and high strength steels. *International Journal of Fatigue* 25 (5): 487–495.

Stacey, A., Barthelemy, J.Y., Leggatt, R.H., and Ainsworth, R.A. (2000). Incorporation of residual stresses into the SINTAP defect assessment procedure. *Engineering Fracture Mechanics* 67 (6): 573–611.

Stacey, A., Sharp, J.V., and King, R.N. (1996). High strength steels used in offshore installations. *International Conference of OMAE*. Stout, Doty, Epstein, and Somers, Weldability of steels (1978). New York: Welding Research Council.

Teng, T.L., Fung, C.P., Chang, P.H., and Yang, W.C. (2001). Analysis of residual stresses and distortions in T-joint fillet welds. *International Journal of Pressure Vessels and Piping* 78 (8): 523–528.

Ueda, Y. and Nakacha, K. (1982). Simplifying methods for analysis of transient and residual stresses and deformations due to multipass welding. *Transactions of JWRI* 11 (1): 95–103.

Ueda, Y., Takahashi, E., Fukuda, K. et al. (1976). Transient and residual stresses from multipass welding in very thick plates and their reduction from stress relief annealing. *Transactions of JWRI* 5 (2): 179–187.

Ueda, Y., Wang, J., Murakawa, H., and Yuan, M.G. (1993). Three dimensional numerical simulation of various thermo-mechanical processes by FEM (Report I). *Transactions of JWRI* 21 (2): 111–117.

Wang, J., Ueda, Y., Murakawa, H. et al. (1996). Improvement in numerical accuracy and stability of 3-D FEM analysis in welding. *Welding Journal* 75 (4): 129–134.

Webster, G.A. and Wimpory, R.C. (2001). Non-destructive measurement of residual stress by neutron diffraction. Journal of Materials Processing Technology 117 (3): 395–399.

Willms, R. (2009). High strength steel for steel constructions. *The Nordic Steel Construction Conference (NSCC2009)*. Sweden.

Wingerde, A.M.v., Packer, J.F., and Wardenier, J. (1995). Criteria for the fatigue assessment of hollow structural section connections. *Journal of Construction Steel Research* 35 (1995): 71–115.

Withers, P.J. (2007). Mapping residual stress and internal stress in materials by neutron diffraction. *C.R. Physique* 8 (2007): 806–820.

Wu, A., Ma, N.X., Murakawa, H., and Ueda, Y. (1996). Effects of welding procedures on residual stresses in T-joints. *Transactions of JWRI* 25 (4): 81–89.

Zhao, X.L., Herion, S., Puthli, J.A. et al. (2000). Design guide for circular and rectangular hollow section welded joints under fatigue loading. CIDECT, TUV.

Index

Note: Page numbers in *italics* refer to figures and those in **bold** to tables.

a

ABAQUS finite element modeling package 64, 67, 132
American Welding Society (AWS)
  hot-spot stress definition 34
  standards **13**, 100, 101, 144
  weldability definition 15
arc welding 13–15
  electrodes 4, 14
  flux-cored arc welding 14, 77, 100, 101, 130
  shielded metal arc welding 14, 39, 40, 72, 77
  welding speed controlled by welder 130
ASTM standards
  brace plate cutting 38, 47, 52–54, 61
  calibration tests 45, *46*, **46**, 107
  hole drilling method for residual stress investigation 24, 28, 37, 38–39
  materials specifications **13**
austenite, transformation from and to other structures 9, 10, 17, 18
AWS *see* American Welding Society
axial load (AX)
  hot-spot stress 35
  nominal stress 33
  stress concentration factors 153, 154, *155*, 158–160, *160*, 164

b

bainite, transformation to ferrite and cementite 10
b/c *see* brace width to chord width ratio
boundary conditions, numerical modeling of plate-to-plate welded joints 90, *90*, 91, *91*
box hollow section (BHS) joints
  CIDECT design guide 35, 161
  classification and parameters 31, *32*
  experimental investigations, stress concentration factors 153–164
  experimental residual stress investigation 99–123
    analysis and discussion 118–123
    box section fabrication 104
    chord edge effects 118, 120, *120*, 123
    corner effects 120–121
    fabrication procedure 101–105, *102*, *103*, *104*, *105*
    joint intersection fabrication 104–105
    preheating effect analysis 118–119
    residual stress computation procedures 107–108, *109*
    stress variation in depths 121–122, *121–122*, **122–123**
    test results 109–118
    test setup and procedure 105–107, *105*, *106*
    welding design 101–103

Welded High Strength Steel Structures: Welding Effects and Fatigue Performance, First Edition. Jin Jiang.

fatigue analysis 31–35
numerical modeling of residual stress 125–151, 181–194
arc touch movement 130, *131*
fully coupled thermo-mechanical analysis 125, 126, **131**, 136–141
heat source modeling 129, 175–179
methods 125–126
modeling procedure 126–132, *127*, *128*, *132*
parametric study 141–150
plate-to-plate modeling joint comparison **131**
preheating 129–130, 132–136
source code 175–177, 181–194
thermal interactions 129–130, *131*
stress (fatigue) analysis 32–35
versus plate-to-plate joints 99, 126
brace width to chord width ratio (b/c), numerical modeling of box welded joints parametric study **142**, 145–147, *146*, 149

C

calibration coefficients, hole-drilling methods 24, 38, 46, **46**
calibration plates 45
calibration tests
ASTM standards 45, *46*, **46**, 107
residual stress investigations 45–46, *46*
carbon
effects in steel 15–16
interactions with other elements in steel 15–16
carbon equivalent (CE)
weldability equations 16
yield stress relationship *11*, *16*
carbon-equivalent limit (C_{equiv}) 15
CCT *see* continuous cooling transformation diagrams
CE *see* carbon equivalent
CGHAZ *see* coarse-grained heat-affected zone
chord and brace hollow section fabrication for tubular structures 99, 100, 101–102
chord-brace intersection
analysis area for numerical study 126
joint fabrication 104
weld influence on residual stress 140–141
welding speed variation in box joints 130, *131*
weld simulation 127, *127*
CIDECT design guide 35, 161, 164
coarse-grained heat-affected zone (CGHAZ) **12**, 17
codes and standards
American Welding Society **13**, 100, 101, 144
CIDECT design guide 35, 161, 164
HSS applications 12, **13**
see also ASTM standards
coefficient of thermal expansion (CTE), different materials linked together causing stress 21
cold cracks 19
computational costs
balancing with accuracy 150
see also lumping technique
continuous cooling transformation (CCT) diagrams 17, *18*
cooling rate
calculating average rate 6
effect of heat input rate 14
effect of preheating 76, *76*, 92–93, *92*, 118
effects of lumping schemes 94, *94*
effect on weld 4, 15, 18, *18*
effect of welding sequence 96, *96*
effect of welding speed 94, *95*
residual stress relationship 52, *52*
variation across different locations of box joint 134–136, *135*, 150
crack growth rate, residual stress effects 30–31, *31*
CTE *see* coefficient of thermal expansion
cyclic loading
box joints 4, 99, 126, 153
plate-to-plate joints 37
cyclic plastic deformation at weld toes, eliminating tensile residual stresses 29

d

Dearden equivalent, carbon equivalent equation 16
delivery condition of high strength steel 10–11
discretization of plate-to-plate joints, finite element modeling 65, *66*

e

EC3 *see* Eurocode 3
electrodes
  arc welding 4, 14
  used in experimental investigations 39–40, **40**, 100–101, **101**
element birth and death method
  residual stress in box joints **131**, 150
  weld filler simulation 67, *67*, 96
element movement method, weld filler addition modeling 67
element types, modeling box joints versus plate-to-plate joints **131**
energy balance principle, heat conduction through a volume element 23
equivalent carbon content *see* carbon equivalent
Eurocode 3 (EC3) 34, 68, 69, *69*, 70–71, *70*
experimental residual stress investigations
  future research recommendations 168
  overall conclusions 166–167
  overview 24–25, **25**, 165
  plate-to-plate joints 37–61
  welded box T-joints 99–123
external loading (tensile stress)
  residual stress reduction effects 29, *29*
  static tensile testing, plate-to-plate welded joints 54–57

f

fatigue problems in HSS structures 11–12
FCAW *see* flux-cored arc welding
ferrite, transformation from and to austenite 10, 17
FGHAZ *see* fine-grained heat-affected zone
filler material addition in welding, numerical modeling of plate-to-plate joints 27–28, 67–68, 78–79, *78*, *79*, *80*, 96
fine-grained heat-affected zone (FGHAZ) 17
finite element modeling
  ABAQUS package 64, 67, 132
  plate-to-plate welded joints 63–97
    ambient temperature results 81–83, 87
    comparison with testing results **72**, **73**
    discretization of joints 65, *66*
    heat source model 77–78, *77*
    heat transfer analysis 68–70, 72–73
    lumped/lumping techniques 27, 64–66, 78–79, *78*
    lumping schemes 90, *90*, *91*, 93, *94*
    mechanical analysis 70–71, *70*
    mechanical boundary conditions 90, *90*, 91, *91*
    modeling procedures 64–67, *65*, **66**, 76–80
    model validation 71–72, 80–81
    number of weld lumps 93–94, *93*, *94*
    parametric study 87–96
    preheated joint results 83–87, 97
    residual stress results 74–76, *74*, *75*, *76*
    results **71**, 72–76, 81–87, 171–173
    three dimensional models 76–87, 171–173
    two-dimensional models 64–76, 87–96
    weld filler addition 27–28, 67–68, 78–79, *78*, *79*, *80*, 96
    welding sequence effects 90, *90*, *91*, 95–96, *96*
    welding speed effects 94, *95*
flux-cored arc welding (FCAW) 14, 77, 100, 101, 130
Fourier's law of heat conduction 22, 77

fracture mechanics
  distribution of average critical linear elastic fracture mechanics parameter at ambient temperature **12**
  fatigue performance investigation of tubular/box joints 153
  fatigue problems in HSS structures 11–12
  offshore analysis 30
  see also hot-spot stress/strain
fracture toughness property 12, **12**, 18
full element method, weld filler addition modeling 67
fully coupled thermo-mechanical analysis
  box joints
    advantages over sequential coupling 125
    modeling validation 136–137
    residual stress modeling 125, 126, **131**, 136–141
    residual stress results 138–141
    temperature history results 137–138
fusion welding
  processes 13, 14–15, *15*
  residual stress production 63
  see also arc-welding
future research recommendations 168

g

grain size and structure
  cooling rate 14
  high strength steel production methods 11, *11*
  welding effects 14, 17–19, *18*
  welding speed effect 149
  weld zones **12**, 14, 18

h

heat-affected zone (HAZ)
  welding **12**, 14, 15, 17–18, 51, 63, 68, 149
  see also residual stress
heat generation by plastic dissipation 125
heat source modeling, plate-to-plate welded joints 3D analysis 77
heat transfer analysis
  Fourier's law of heat conduction 22
  residual stress generation in the welding process 22–24, *23*
  sequentially-coupled thermal-stress analysis 63, 64, **66**, 68–70, 72–73
heat transfer process
  numerical analysis
    plate-to-plate welded joints 68–69
    preheated box joint welding 132–136, *133*, *134*, *135*, 150
high-strength steel (HSS)
  applications 1, *2*
  box joints versus plate-plate joints 99
  mild steel comparisons 1, 3, 9, 12, 30, 31, 37, 71
historical perspectives
  hole drilling methods 24, 38
  production processes for rolled steel products *3*
hole drilling methods for residual stress investigation
  ASTM standard procedure 24, 37, 38–39
  features 24, **25**
  modification for plate-to-plate welded joint investigation 43, *44*
  welded box T-joints 105–106
hollow sections, chord and brace fabrication for tubular structures 99, 100, 101–102
hot cracks, welding 19, *19*
hot-spot stress/strain
  definitions 34
  stress concentration factors 35, 54–56, *55*, **57**, 153
  tubular joints 34–35, *34*, 153, 154, 156, 159, 166
HSS *see* high-strength steel
hydrogen-induced cracks 19

i

ICHAZ *see* inter-critical heat-affected zone
in-plane bending (IPB)

hot-spot stress 35
residual stress measurement 39
stress concentration factors 153, 154, *155*, 158–160, *160*
Instron Model 8506 Dynamic Materials Testing System 54–57
inter-critical heat-affected zone (ICHAZ) 17
International Institute of Welding (IIW)
carbon equivalent *11*, 16
hot-spot definition 34
IPB *see* in-plane bending

j

joint corners
effects on experimental residual stress results 120–121
stress concentration factors of box joints 163, 164
welding speed variation 130, *131*
joint intersection fabrication, residual stress investigation of welded box T-joints 104–105

l

LB-70L electrode, use in experimental investigations 39–40, **40**
literature review 9–35
load types
stress concentration factors 153
see also axial load; in-plane bending; out-of-plane bending
local stress 24, 29, 33, 34
lumping techniques
numerical modeling of multi-pass welding 27
box joint welding 127, **131**, 150
plate-to-plate welding 64–66, 78–79, *78*, 167

m

manganese, effects in steel 15
manual arc welding process *see* shielded metal arc welding
martensite
cooling rate effects 18
transformation to and from other structures 9, 10, *10*, 18
mean stress, fatigue strength and external loading effects 29
mechanical analysis, numerical modeling of plate-to-plate welded joints 70–71, *70*
mechanical properties of steel, welding effects 18–19, *20*
meshing, numerical analysis of weld 64, *66*, 132, *133*
microstructure of steel 9
modeling *see* finite element modeling; numerical modeling
multi-pass welding
numerical modeling
ambient temperature versus preheating 76
direction of passes 78, 79
lumping techniques 27, 64–66, 78–79, *78*, 97, 127, *131*, 150, 167
preheating effect on later passes 96

n

neutron diffraction method for residual stress investigations 24–25, **25**, 28, 37, 125, 168
nominal stress
plate-to-plate joints 58, *59*, *60*
tubular/box joints 33, *34*, 35, 159
notch stress, tubular joints 35
numerical modeling of pure heat transfer, preheated box joint welding 132–136, *133*, *134*, *135*
numerical modeling of residual stress 26–28, *26*, **27**
plate-to-plate welded joints 63–97, *78*, 167
2D modeling 64–76, 87–96
3D modeling 76–87, 171–173
lumping techniques 64–66, 78–79, *78*, 167
welded box joints 125–151
2D modeling 125–126

3D modeling 126–151, 171–179, 181–194
lumping techniques 97, 123, 127, *131*, 150
parametric study 141–150
versus plate-to-plate joints **131**
see also finite element modeling
numerical modeling of welded joints 5, 6
future research recommendations 168
overall conclusions 167–168
overview 165–166

o

offshore structures, fatigue and fracture 11–12
OK Tubrod 15.09 electrode 100–101
out-of-plane bending (OPB), stress concentration factors 153, 154, *155*, 158–160, *161*

p

parametric studies
2D numerical modeling of plate-to-plate welded joints 87–96
boundary conditions 91, *91*
lumping schemes 93–94, *93*, *94*
preheating temperature 91–93, *92*
welding sequence 95–96, *96*
welding speed 94, *95*
numerical modeling of box welded joints 141–150
b/c ratio (brace width to chord width) **142**, 145–147, *146*, 149
joint angle 142–145, *142*, **142**, **143**
preheating temperature **142**, 143–145, *144*, 145–146, *145*, *146*, 147–149, *148*
welding speed **142**, 147–150, *149*
weld starting location 142–143, *142*, **142**, *143*, 146–147, *147*
pearlite (ferrite plus cementite), transformation from and to other structures 10, 17, 18
phase transformation, welding 16–18, *17*
phosphorus, effects in steel 16
plastic dissipation, heat generation 125
plastic strain, hot cracks 19
plate-to-plate welded joints
experimental residual stress investigation 37–61
brace plate cutting effects 52–54, *53*, *54*, 61
brace plate cutting procedure 47, *47*, *48*
calibration test 45–46, *46*, **46**
distribution along weld toe 49, *49*, *50*
experimental results 47–54, **48**
hole drilling method 38–39, 43
joint angle effects 50–51, *51*, *52*, 61
measurement procedure 46–47
plate thickness effects 51–52, 61
preheating effects 49–50, *50*, *52*, *54*, 60
setup 43, *44*
specimen specifications 39–41
strain gauge locations 44–45, *45*
welding specifications 41–43
numerical modeling 63–97
parametric study 87–96
three dimensional finite element analysis 76–87, 171–173
two dimensional finite element analysis 64–76, 87–96
static tensile testing 54–57
versus box hollow section joints 99, 126
welding profile/geometry *41*, **41**
preheating of weld joints
experimental investigation of plate-to-plate joints
effects on residual stress 49–50, *50*, *52*, *54*, 60
hole drilling stress test method 41–42, *42*
experimental investigation of welded box T-joints
analysis/conclusions 118–119, 123
results 109–118, *117*, *119*

specimen fabrication 101, 104, *104*, 105
test setup 106–107, *107*
numerical modeling of box joints 129–130
allowing for cooling/heat transfer 132–136, *133*, *134*, *135*
parametric study **142**, 143–146, *144*, *145*, *146*, 147–149, *148*
numerical modeling of plate-to-plate joints
2D modeling 69, 71–76, 92–93, *93*
3D modeling 76, *78*, *81*
temperature effects, 2D numerical modeling of plate-to-plate welded joints 91–93, *92*
production methods of high strength steel 9–11, *10*, *11*
pure heat transfer, modeling preheated box joint welding 132–136, *133*, *134*, *135*

q

quenching and tempering (QT), procedure effects on steel microstructure 9, 10, *11*

r

research background 1–8
residual stress 20–31
box joints versus plate-plate joints 99
computation procedures 107–108, *109*
eliminated by cyclic plastic deformation at weld toes 29
experimental investigations 24–25, **25**
plate-to-plate joints 37–61
welded box T-joints 99–123, 153–164
exploration of effects 28–31
generation by welding 22–24, *23*
numerical modeling 26–28, *26*, **27**
plate-to-plate joints 63–97
welded box joints 125–151
welded box T/Y-joints 125–151
origins 20–22, *21*
reduced by high applied stresses in box joints 164
removal by thermal stress relief 29
sign conventions *110*
types by magnitude 22, *22*
residual stress factor (RSF), calculations 58–60, *59*, *60*
RQT701 HSS plate for use in experimental investigations, specifications 39–40, **40**
RS-200 milling guide, hole drilling method for residual stress measurement 43, *44*

s

SCFs *see* stress concentration factors
SCHAZ *see* sub-critical heat-affected zone
semi-automatic arc welding process *see* flux-cored arc welding
sequentially coupled thermo-mechanical analysis
numerical modeling of plate-to-plate joints 64, *66*, 71, 96, **131**, 165, 167
versus fully coupled 125, **131**
shielded metal arc welding (SMAW) 14, 39, 40, 72, 77
sign conventions, residual stress *110*
silicon, effects in steel 15–16
SMAW *see* shielded metal arc welding
S–N curve approach
fatigue performance investigation of tubular/box joints 32–34, *33*, 153
fatigue performance of welded joints dependent on mean strength 29
solid-phase welding 13, 14
Sony Center, Berlin 1, *2*
standards *see* codes and standards
static testing
box joints *5*, 153–164
equipment and procedures 154–159
residual stress effects 162–163
test results 159–161
plate-to-plate joints *5*, 54–57

equipment and procedures 54–56, *55*, *56*, *57*
test results 56, **57**
strain gauges
rosette design for hole drilling method of residual stress testing 34, 37, *39*, 44–45, *45*, 105–106
schemes for residual stress measurement, box T-joints 106–107, *107*, *108*
schemes for stress concentration factors assessment on box T-joints 156–157, *156*, *157*, *158*
strain hardening, reduced in high-strength steel 3, 9, 37
strength of steel
wall thickness and weight relationships *3*
see also tensile strength; toughness properties; weldability
stress concentration factors (SCFs)
different load types 153
hot-spot stresses 35, 54–56, *55*, **57**, 153
influences of residual stress 57–60, *59*, *60*, 162–163
plate-to-plate joint residual stress effects *55*, 56, 57–60, **57**
static test on large-scale preheated box joints 153–164
axial load 153, 154, *155*, 158–160, *160*, 164
comparison with CIDET guide 161, **161**, 164
corners 163, 164
in-plane and out-of-plane bending 153, 154, *155*, 158–160, *160*
residual stress effects 162–163, *162*, *163*, *164*
results 159–160, *160–161*
setup and specimens 154–156
strain gauge schemes 156–157, *156*, *157*, *158*
test procedure 158–159
strain gauges *55*
stress-life *see* (S–N) curve approach
sub-critical heat-affected zone (SCHAZ) 17
sulfur, effects in steel 16
surface quality of steel, effects of carbon and additional elements 15–16

t

temperature (thermal) history 10, *10*, 68–70, 73, *74*
tempering of high-strength steel 9, 10, *10*, *11*
tensile strength
steel composition and structure 9, 11, 15
weldability relationship 12
tensile stress
hot cracks 19
residual stress reduction 29, *29*
thermally generated stresses 21
thermo-mechanical analysis of residual stress 125
see also fully coupled thermo-mechanical analysis; sequentially coupled thermo-mechanical analysis
thermo-mechanical control process (TMCP), HSS production 11
three dimensional numerical modeling
residual stress in box joints 126–151, 175–179, 181–194
residual stress in plate-to-plate welded joints 76–87, 171–173
source code of box joint modeling 181–194
source code of heat source 175–179
TMCP *see* thermo-mechanical control process
toughness properties 11, 12, **12**, 18
tubular structures *see* box hollow section joints
two dimensional numerical modeling of residual stress
box joints 125–126
finite element modeling of plate-to-plate joints 64–76
parametric study of plate-to-plate joints 87–96

u

ultrasonic analysis **25**, 105

w

weldability of steel 15–16, *16*
weld filler addition processes, numerical modeling 27–28, *66*, 67–68, *67*, 78–79, *78*, *79*, *80*, 96
welding **12**, 13–20
  couplings in process 26, *26*, **27**
  flux-cored arc welding 14, 77, 100, 101, 130
  heat-affected zone **12**, 14, 15, 17–18, 51, 63, 68, 149
  heat transfer analysis 22–23, *23*, 64, **66**, 68–70, 72–73, 132–136, 150
  high strength steel compared to mild steel 4
  shielded metal arc welding 14, 39, 40, 72, 77
  three metallurgical zones 14
welding speed
  controlled by welder in flux-cored arc welding 130
  grain structure 18–19, *18*
  numerical modeling of box joints **142**, 147–150, *149*
  numerical modeling of plate-to-plate welded joints 94, *95*
  variation around box joint sides and corners 130, *131*

x

X-ray diffraction method for residual stress investigation 24, **25**, 37, 125

z

zones of transformation in welding 17–18
  see also heat-affected zone

Printed in the USA/Agawam, MA
December 19, 2023

857502.008